Guidebook to stereochemistry

Guidebook to stereochemistry

F. D. Gunstone
University of St. Andrews, Scotland

With a foreword by J. I. G. Cadogan

Longman
London and New York

LONGMAN GROUP LIMITED
London

Associated companies, branches and representatives throughout the world

Published in the United States of America by Longman Inc. New York

First published 1975

Library of Congress Cataloging in Publication Data

Gunstone, F. D.
Guidebook to stereochemistry.

Bibliography: p.
Includes index.
1. Stereochemistry. I. Title.
QD481.G87 541′.223 75–12762
ISBN 0–582–44170–6

Set in Monophoto Times New Roman
and printed in Great Britain
by William Clowes & Sons, Limited
London, Beccles and Colchester

Contents

Acknowledgements

We are grateful to the following for permission to reproduce copyright material:
Figs 2 and 3 from Natta and Farina, *Stereochimica* published by Arnoldo Mondadori Editore, Milan; English translation by Andrew Dempster, *Stereochemistry* published by Longman.

Foreword

A century has passed since van't Hoff and Le Bel gave stereochemistry its surging start with their statement that molecules have three dimensional structures. Nowadays it is a statement of the obvious to say that the importance of this concept to the development of chemistry, and of organic chemistry in particular, cannot be over emphasised, because these ideas, coupled with the experimental advance associated with the determination of accurate molecular weights by the vapour density method, enabled organic chemistry to become a science. Since then stereochemistry has grown to occupy a pivotal role in organic, organometallic and inorganic chemistry, and nowhere is it more important than in the understanding of the molecular processes and macromolecular structures vital to biology.

Professor Gunstone's timely little "Guidebook" is based on his long contact as a teacher with generations of students of elementary chemistry, geology, biology and medicine as well as with those of Honours Degree Chemistry. The author has chosen to scatter problems, questions and exercises liberally through his text, so that the reader is almost forced to work through examples as part of the learning process, the aim being that, having done this, the student will be familiar with all the basic concepts and terminology and will understand why they are important not only in organic chemistry, but also in related fields such as organo-metallic chemistry, polymer chemistry and enzymology.

The result is a clear summary and explanation of the essentials of the subject which will be useful to the beginner as much as to more accomplished students and their teachers.

J. I. G. Cadogan
September 1974

Author's preface

Most chemists agree that the theory of molecular configuration published independently and almost simultaneously in 1874 by Joseph Achille Le Bel and Jacobus Henricus van't Hoff was a seminal concept in the deeper understanding of molecular structure and in the ideas of stereochemistry which grew from the proposals of these two chemists. It is particularly appropriate that this book should be written exactly one hundred years later.

Structure has always been of fundamental importance to chemists and this attitude has spread to workers in other scientific disciplines as they have become attuned to thinking at the molecular level. A century after the proposals of Le Bel and van't Hoff the representations used by chemists and others to indicate molecular architecture have become more refined and more meaningful. Since molecular assemblies exist in a three-dimensional framework structures must also be interpreted in this way. This is best achieved with some form of molecular kit, though two-dimensional representations sometimes suffice and students of chemistry must learn to interpret two-dimensional figures in three-dimensional terms. The reader will gain more from this book if he has a set of atomic models from which he can construct molecular arrays to check statements made in the text and to carry out the suggested exercises.

This Guidebook is designed to lay a foundation in the understanding of stereochemistry and separate chapters deal with *cis-trans*-isomerism, enantiomerism, conformation, and dynamic stereochemistry. Interspersed through the text are questions and exercises and to get the most out of this book these should be undertaken as the text is read. Answers are given at the end of each chapter. These are an important feature of the book and some information is only contained in the questions and their answers. The active participation of the reader is essential if he is to be able to *use* the concepts of stereochemistry after he has read the book. When this study has been completed the reader should be able to:

(i) Recognise structures of compounds which display *cis-trans*-isomerism and enantiomerism.

(ii) Formulate stereoisomers as sawhorse, Fischer, and Newman projections and interconvert these.

(iii) Assign the symbols *E* and *Z* to *cis-trans*-isomers and symbols *R* and *S* to chiral centres.
(iv) Recognise the differences in physical and chemical properties (if any) between stereoisomers.
(v) Discuss and explain conformation in simple acyclic compounds and in compounds derived from cyclohexane.
(vi) Formulate addition, elimination, and substitution reactions in configurational terms and discuss configurational relationships between substrate and product.
(vii) Use correctly the terms: chiral, achiral, and prochiral; chiral centre, chiral axis, and chiral plane; chair, boat, and twist conformation; *cis-trans*-isomerism; enantiomer, diastereoisomer, meso compound, and racemic form; *R*, *S*, *E*, and *Z*; sawhorse, Newman and Fischer projections; conformation, eclipsed, staggered, synperiplanar, antiperiplanar, synclinal, and anticlinal.

I acknowledge the help I have received from many sources during the preparation of this manuscript including my colleagues Gordon Harris, Douglas Lloyd, Ray Mackie, and David Smith. Dr R. S. Cahn and Professor J. I. G. Cadogan read the manuscript and made helpful comments which contributed to its improvement. I thank Mrs W. Pogorzelec for her careful and patient typing.

F. D. Gunstone
St. Andrews, 1974

1

Introduction

1.1 Structural formulae

Chemistry is primarily a molecular science. This means that chemists seek to explain observable phenomena in terms of the molecular structure of the changing material. Faced with a problem, a chemist will often ask "what is the structure of the material?" Molecular structures are models or shorthand descriptions of molecules which convey considerable information to those who understand these things. Molecular structure provides a basis for understanding observed properties, for predicting new properties not yet reported, for suggesting methods of synthesis, and, if the compound is a natural product, for speculating about its method of production (biosynthesis) and its reactions in living systems (metabolism). The importance of structure is reflected in the time spent by chemists in determining structure and in the effort devoted to devising and developing new procedures – both chemical and physical (mainly spectroscopic) – for furthering such studies.

With the passage of time chemical structures have become increasingly sophisticated and, we believe, more exact. Every compound has its own distinct structure and if two compounds differ in only a single property, though identical in all other respects, then their molecules are different in some way and this must be reflected in their differing structures. This fundamental belief has, at certain times, led to the introduction of completely new structural concepts as, for example, the suggestion made independently by Le Bel and by van't Hoff in 1874 that molecules exist in a three-dimensional form and the later development of ideas of conformation and of resonance.

The importance of structure and its continuous refinement is well illustrated in the case of the important monosaccharide glucose which has been represented at various times by the structures **1** to **7**. In common with other carbohydrates, glucose was first represented by structure **1** showing it to consist of the three elements, carbon, hydrogen, and oxygen

$(CH_2O)_n$ (**1**)

$C_6H_{12}O_6$ (**2**)

CHO
CHOH
CHOH
CHOH
CHOH
CH_2OH
(**3**)

CHO
H—C—OH
HO—C—H
H—C—OH
H—C—OH
CH_2OH
(**4**)

HO H
C
H—C—OH
HO—C—H
H—C
H—C—OH
CH_2OH
O
(**5**)

HO H
C
H—C—OH
HO—C—H
H—C—OH
H—C
CH_2OH
O
(**6**)

H, CH_2OH, H, O, HO, HO, H, H, OH, H, OH
(**7**)

in the proportion 1:2:1. When its molecular weight became known it could be more correctly represented as **2** but this is not a satisfactory representation of glucose for at least two reasons. First of all it is not very informative; it does not, for example, show the order of the 24 constituent atoms: secondly it does not distinguish glucose from many other substances also correctly written as $C_6H_{12}O_6$. Structure **3** indicates which atoms are actually linked together whilst **4** is a two-dimensional representation of a three-dimensional molecule and follows the development of the idea of asymmetry or chirality discussed in Chapter 3. At a still later date it was recognised that structure **4** was inconsistent with those properties of glucose in which it did not behave like a typical aldehyde and in which it reacted like a pentahydroxy compound. The five- and six-membered hemi-acetal ring structures (**5** and **6**) then became more acceptable because they provided a better description of the properties of glucose. At a still later date it was accepted that the cyclic molecule was not planar and that one form of glucose existed mainly in the conformation shown as **7**. The simpler structures of glucose are still used but the reader is then expected to know that this is only for convenience and that they are to be understood as abbreviated forms of the most accurate structural representation.

Early in the history of organic chemistry it was recognised that more than one compound could be represented by the same molecular formula and these compounds were described as isomers. In this book we shall be concerned with the subtle kind of isomerism called stereoisomerism which can be understood only in three dimensional terms.

For example, tartaric acid has been known for centuries, being readily obtained from a crystalline deposit which separated in wine vats in the form of a potassium salt. About 1819 a second acid obtained during the crystallisation of the above potassium salt was called racemic acid (*racemus*, bunch of grapes). The two acids have the same chemical properties and both can be represented as HOOCCH(OH)CH(OH)COOH but they differ in such physical properties as their crystalline form, melting point, and water-solubility. In 1853 Pasteur made yet another compound (mesotartaric acid) differing from the others but having the same structure when represented in the linear form shown above. The problems raised by these observations are discussed in Chapter 3.

Under appropriate conditions benzene can be chlorinated to give benzene hexachloride (**8**). This compound exists in several isomeric forms but only one isomer shows the marked insecticidal action which makes this compound important. Many examples are now known of pharmaceuticals, flavours, and insect attractants, where only one isomer produces the desired effect.

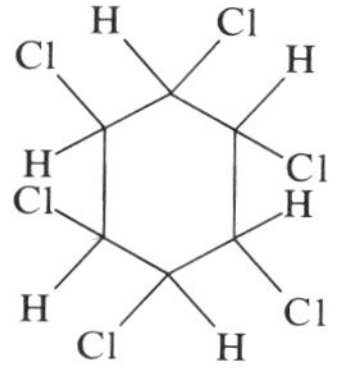

(**8**)

1.2 Enantiomers and diastereoisomers

Two or more compounds whose structures differ only in the three-dimensional arrangement of their constituent atoms are said to be stereoisomers and within this group there are two subgroups: enantiomers and diastereoisomers. *Two stereoisomers which differ only in their ability to rotate plane polarised light in an equal and opposite direction are enantiomers.*† *All other stereoisomeric compounds, whether they are able to rotate plane polarised light or not, are diastereoisomers.* Enantiomeric pairs are mirror images, diastereoisomers are not.

This is a wider, but more satisfactory, classification of stereoisomers than was used in the past. Formerly the terms enantiomer and diastereoisomer were applied only to optically active (asymmetric, chiral) compounds and geometric (*cis trans*) isomers were not considered as diastereoisomers (see, for example, E. L. Eliel, 1971).

This new classification is summarised in the block diagram which is to be interpreted as follows:

(i) stereoisomers are of two distinct kinds: enantiomers and diastereoisomers,
(ii) enantiomers result from chirality only: diastereoisomers result from chirality or *cis-trans*-isomerism,
(iii) chiral systems may be enantiomeric or diastereoisomeric: *cis-trans*-isomers are diastereoisomeric and never enantiomeric (unless they also contain a chiral system).

<table>
<tr><td colspan="3">stereoisomers</td></tr>
<tr><td>enantiomers</td><td colspan="2">diastereoisomers</td></tr>
<tr><td colspan="2">chirality</td><td>cis-trans-isomerism</td></tr>
</table>

1.3 The sequence rules

On several occasions in later chapters it will be necessary to designate the *order* of groups (ligands) attached to a particular atom. This is now done with the help of so-called Sequence Rules and it is useful to explain these now so that they can be used later without interruption of the argument then in progress. We owe these ideas mainly to Cahn and Ingold (1951) and to Cahn, Ingold, and Prelog (1956, 1966).

It is not sufficient that students *understand* these rules, they must be able to *use* them and therefore a number of examples are incorporated and the reader is recommended to test his mastery of the rules. There are

† Such compounds are chiral and show other differences in the presence of chiral reagents. They may also smell differently.

several Sequence Rules, some of which are required only for advanced examples. We shall confine ourselves only to the simpler cases.

```
     C   P   C
     |   |   |
B—A—X—A—B
     |   |   |
     D   Q   E
```

The Sequence Rules allow us to put the various ligands attached to a particular atom (such as **X**) in an agreed order of precedence by comparing the ligands according to agreed rules (the Sequence Rules). Comparisons are made between the *atoms* attached to **X** (P, Q, A, and A in the above figure). If this is not sufficient to settle the order of precedence then the atoms attached to one of the equivalent atoms (B, C, and D in the structure shown) are compared in hierarchical order with the atoms attached to the other equivalent atom also in hierarchical order (B, C, and E). If this does not settle the issue then further sets of atoms come under consideration until a decision can be made.

The rules we shall use state (i) that ligands (as defined above) are arranged in order of *decreasing atomic number*, (ii) that isotopes (when present) are arranged in *decreasing mass number*, and (iii) that when required a lone pair of electrons is placed below hydrogen (^1H) in the precedence order. These rules are applied successively as far as is necessary to reach a decision on priorities, i.e. (ii) only if (i) does not give the answer, and so on. The only other thing we need to know is how to treat unsaturated systems but first let us test our understanding of these simple rules.

Q1.1 Place the following sets of ligands in order of precedence (starting with the highest):

(i) hydrogen, bromine, fluorine, and chlorine;
(ii) hydrogen, deuterium, chlorine, and methyl (CH_3);
(iii) $—OCH_3$, $—NHCH_3$, $—SO_3H$, $—Cl$;
(iv) ethyl (C_2H_5), methyl (CH_3), hydroxyl (OH), and hydrogen;
(v) $—OCH_2C(CH_3)_3$, $—OCH_2CH_2Cl$, $—OCH_3$, $—H$.

Ligands containing unsaturation have to be handled with the help of what are known as "ghost" or "phantom" atoms which are always shown in brackets. Before application of the Sequence Rules the system —A=B, for example, is rewritten as a single-bond structure in which A, instead of being doubly bonded to one atom of B, is singly bonded to two atoms of B. Similarly B must be considered to be singly bonded to two atoms of A so that:

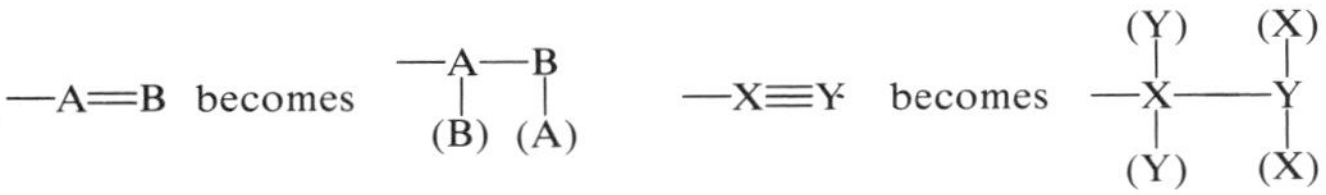

—A=B becomes —A—B with (B) and (A) attached below; —X≡Y becomes —X—Y with (Y) above and below X and (X) above and below Y

The revised structures contain no multiple bonds and instead a number of ghost atoms.

Q1.2 Represent the following ligands in the form required for application of the Sequence Rule:

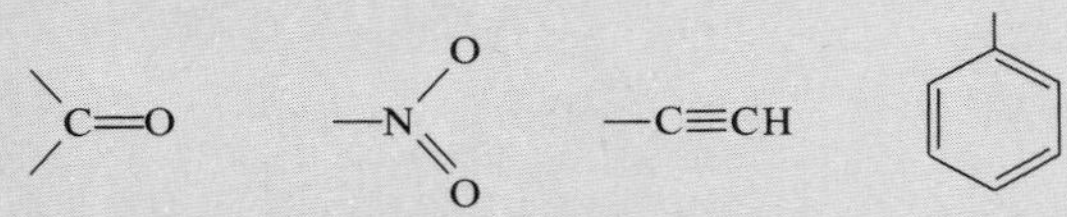

>C=O (carbonyl) —NO_2 (nitro) —C≡CH (ethynyl) (phenyl)

Answers

A1.1 (i) Br, Cl, F, H. This follows from the first rule – the atoms are listed in order of decreasing atomic number.

(ii) Cl, CH_3, D, H. In considering the methyl group only the first atom (C) need be considered in this example.

(iii) Cl$>SO_3H>OCH_3>NHCH_3$. This precedence order is easily settled on the basis of the first atom Cl>S>O>N.

(iv) OH, C_2H_5, CH_3, H. It is easy to see that O>C>H, but this leaves the ordering of C_2H_5 and CH_3 to be settled. Since it cannot be decided on the basis of the first atom (both C) then the next set of atoms have to be considered in order of precedence, i.e. C(HHH) has to be compared with C(CHH). Notice that the atoms in brackets are themselves listed in order of decreasing precedence. These are compared in turn and the first comparison (H and C) is enough to show that $C_2H_5>CH_3$.

(v) $OCH_2CH_2Cl>OCH_2C(CH_3)_3>OCH_3>H$.
Notice that O C C Cl takes precedence over O C C C.

A1.2

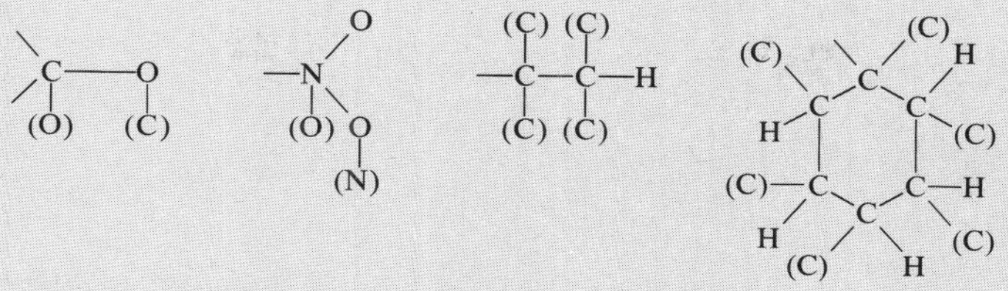

Notice that the phenyl group would be written in the same way starting with the alternative Kekule formulation.

2

Cis-trans-isomerism

2.1 Two examples of historical significance

When malic acid (2-hydroxybutanedioic acid) is heated it is dehydrated as indicated:

$$\underset{\text{malic acid}}{HO_2CCH(OH)CH_2CO_2H} \xrightarrow{-H_2O} HO_2CCH{=}CHCO_2H \xrightarrow{H_2} \underset{\text{succinic acid}}{HO_2CCH_2CH_2CO_2H}$$

This representation, however, is not entirely satisfactory because the product is a mixture of two compounds designated fumaric and maleic acids.† Both have the molecular formula $C_4H_4O_4$, both are reduced to succinic acid (butanedioic acid), and they show other similar properties. Nevertheless they also show some distinctive properties and since they are clearly two different compounds they have to be represented by two different structures.

Many fatty compounds contain oleic acid as a major component. Its structure (**1**) was first correctly reported by Meyer and Jacobsen (1893).

$$CH_3(CH_2)_7CH{=}CH(CH_2)_7CO_2H$$
(**1**)

† The actual product is a mixture of fumaric acid and maleic anhydride but on treatment with water the latter becomes maleic acid.

Under certain conditions this low-melting solid (mp 13° when pure) is converted to a higher melting form designated elaidic acid (mp 45°). Oleic and elaidic acids have similar but not identical properties and both are equally well represented by structure (**1**). It follows that structure (**1**) is not entirely adequate since it cannot describe two acids which however similar are yet different compounds.

These are two early examples of stereoisomeric compounds which are now discussed in terms of *cis-trans*-isomerism.

It is important to realise that experimental observation and theoretical explanation are closely intertwined. Observations such as those above showed the limitation of the structural descriptions then employed. Eventually theories and structures were improved to take account of these unexpected observations, leading, sometimes, to prediction of related phenomena which were observed experimentally only at a later date. Always, or almost always, observation preceded the generalised description of the phenomena which we call "theory". The stereoisomerism illustrated in the two examples discussed above can now be understood in terms of the shape of sp^2 hybridised orbitals but our knowledge of the nature of the so-called double bond derived, in the first instance, from experimental observations such as those described above and not from theoretical prediction.

2.2 The structural requirements for *cis-trans*-isomerism

When two atoms are linked by a single bond the barrier to free rotation about this bond is quite small. It is, nevertheless, real and important and is discussed at length in Chapter 4. On the other hand, when two atoms are multiply bonded then the barrier to free rotation is very much higher and where such compounds exist in two forms the passage from one to the other requires so much energy that each isomer has a stable and independent existence. In short, free rotation about a carbon to carbon double bond cannot occur without disruption of the bond.

In an alkene the two olefinic carbon atoms are sp^2 hybridised and the three sp^2-bonding orbitals lie on one plane. The additional bond formed by overlap of the p_z orbitals greatly restricts the free rotation about the bonds uniting the two olefinic carbon atoms (Fig. 1). The two olefinic carbon atoms and the four attached atoms lie in one plane and, depending on

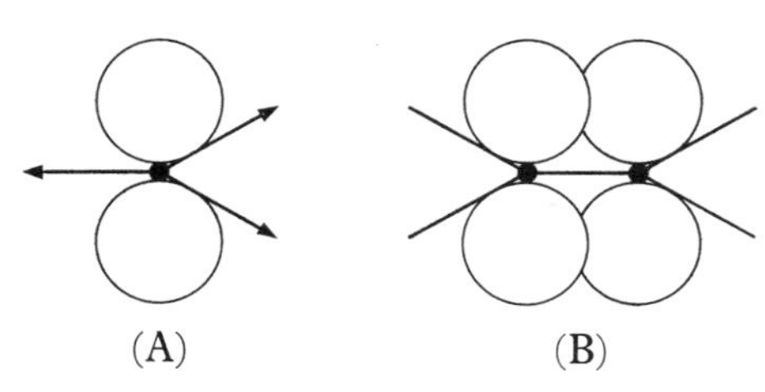

Fig. 1 (A) sp^2-Hybridised carbon atom showing p_z orbital as spheres and three sp^2 hybridised orbitals as lines in one plane and at 120°.
(B) Free rotation in an alkene is restricted because there must be maximum overlap of the two atomic p_z orbitals for maximum stability of the molecular π-orbitals.

the nature of the attached groups, the alkene may exist in more than one form. This type of stereoisomerism is designated *cis-trans*-isomerism. If the alkene is designated as $abC{=}Ccd$ then it will exist in two forms so long as $a \neq b$ and $c \neq d$. It does not matter whether a and/or b are the same as c and/or d.

Q2.1 Examine the structures **2–5** and, if possible, make models of these structures. How many different compounds are represented? Remember that the symbols represent atoms which lie in one plane and that rotation about the C=C bond only occurs under special circumstances which can be neglected for the present.

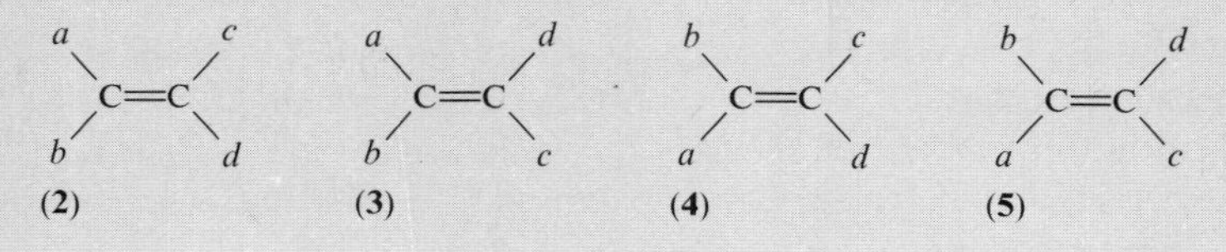

It follows, if we are to distinguish the two "sides" of the double bond, that groups a and b must differ from each other and groups c and d must also differ from each other.

Q2.2 Which of the structures **6–9** represent compounds able to show *cis-trans*-isomerism? Formulate the stereoisomer in these cases and state whether the stereoisomeric pair is correctly described as enantiomeric or diastereoisomeric (see Chapter 1, section 2).

(*a*)(*b*)C=C(*d*)(*c*) **(6)** (*a*)(*a*)C=C(*b*)(*c*) **(7)** (*a*)(*b*)C=C(*a*)(*c*) **(8)** (*a*)(*b*)C=C(*a*)(*b*) **(9)**

2.3 Which compound has which structure?

We now look again at the maleic/fumaric and oleic/elaidic acid pairs. It is easy to formulate structures to correspond to the two pairs of compounds but we then have to face some new questions. Is maleic acid represented by structure **10** or structure **11** and oleic acid by **12** or **13**? Is there a simple

(H)(HO_2C)C=C(H)(CO_2H) **(10)** (H)(HO_2C)C=C(CO_2H)(H) **(11)**

($CH_3(CH_2)_7$)(H)C=C($(CH_2)_7CO_2H$)(H) **(12)** ($CH_3(CH_2)_7$)(H)C=C(H)($(CH_2)_7CO_2H$) **(13)**

way of designating each isomer without having to write out the whole structure each time? It will be convenient to give an answer to the second question first and then to consider the first question. Later in the chapter we will generalise each question and look at the broader issues.

Traditionally the two stereoisomers were designated *cis* and *trans* and these terms are still widely used and are retained in the name now given to this type of stereoisomerism. Previously it was known as geometrical isomerism. The *cis*-isomer is the one which has one or two pairs of identical (or very similar) groups on the same side of the double bond; in the *trans*-isomer the identical (or similar) groups are on opposite sides of the double bond. Structures **10** and **12** therefore represent *cis*-isomers and **11** and **13** represent *trans*-isomers. This system fails when there are no identical (or obviously similar) groups and a more general system of nomenclature is then applied (see p. 13).

The allocation of structures **10** and **11** to the maleic/fumaric acid pair is based on the following observations. Maleic acid (mp 130°) loses water when heated to ~140° to form an anhydride ($C_4H_2O_3$) which on hydration reforms maleic acid. Fumaric acid (mp 270°) is more stable on heating and loses water only at ~275° to form an anhydride ($C_4H_2O_3$) which is identical with that from maleic acid and which on hydration gives *not* fumaric acid but maleic acid. That is to say: maleic acid readily forms an anhydride on heating, fumaric acid forms an anhydride only with difficulty and then it is maleic anhydride. An assignment of structure can now be made on the *assumption* that in the cyclic anhydride the two carbonyl groups must be on the same side of the alkene double bond system.† The alternative would be a highly strained system. Hence maleic acid has its two carbonyl groups on the same side and is the *cis*-isomer and fumaric acid is the *trans*-isomer.

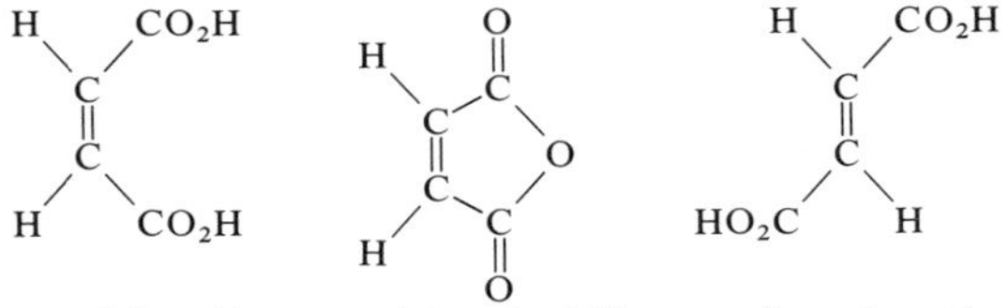

maleic acid: H(CO$_2$H)C=C(CO$_2$H)H (both CO_2H on the same side)

maleic anhydride

fumaric acid: H(CO$_2$H)C=C(H)(HO$_2$C) (the CO_2H groups on opposite sides)

maleic acid   maleic anhydride   fumaric acid

Though originally thought to be the *trans*-isomer, oleic acid is now known to have the *cis*-configuration on the basis of several different lines of argument. One of these involves the use of stereospecific reactions [i.e. reactions of known stereochemical pattern (see Chapter 5)]. The appropriate alkynoic acid (which exists in only a single isomeric form) can be partially reduced to oleic or elaidic acids under reaction conditions now known to furnish a particular stereoisomer.

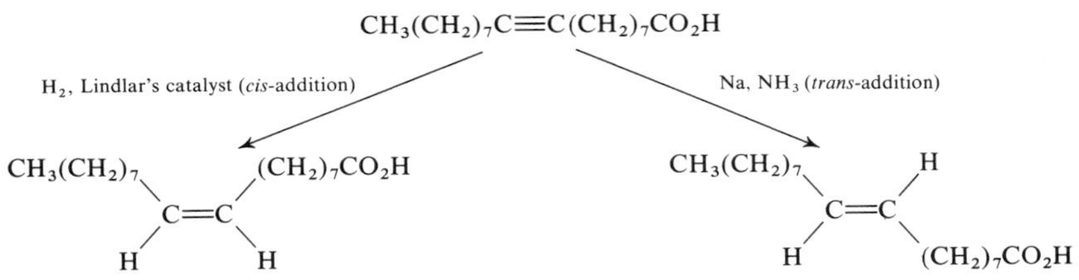

$$CH_3(CH_2)_7C{\equiv}C(CH_2)_7CO_2H$$

H_2, Lindlar's catalyst (*cis*-addition) → oleic acid (*cis*-isomer): $CH_3(CH_2)_7$ and $(CH_2)_7CO_2H$ on the same side of C=C, with H and H

Na, NH_3 (*trans*-addition) → elaidic acid (*trans*-isomer): $CH_3(CH_2)_7$ and $(CH_2)_7CO_2H$ on opposite sides of C=C

oleic acid (*cis*-isomer)   elaidic acid (*trans*-isomer)

† This type of argument must be undertaken with care, especially if the mechanism of the reaction under consideration is not fully known. There are examples where wrong conclusions were drawn through a misinterpretation of the course of reaction (e.g. oximes).

2.4 What happens when the terms *cis* and *trans* cannot be used? The symbols (*E*) and (*Z*)

Difficulty in assigning the terms *cis* and *trans* arises when there is no pair of identical or similar groups. This is now solved by an alternative description which can always be applied. The new terms (*E*) and (*Z*) replace the older terms though it seems likely that *cis* and *trans* will continue to be used to describe alkene isomers for some time yet.

Under the new system the two ligands attached to each olefinic carbon atom are put in order of precedence by application of the Sequence Rules (Chapter 1) and the symbols (*E*) and (*Z*) are then used to describe those isomers in which the ligands of higher precedence are on the opposite (*E* from the German *entgegen*=across) or the same side (*Z* from the German *zusammen*=together). In many cases, *but by no means all*, the old term *cis* corresponds to (*Z*) and the term *trans* corresponds to (*E*). This correspondence should never be used without checking. In order to make sure you understand and can use this new description try the following questions.

Q2.3 Write structures for (i) maleic and fumaric acids and (ii) oleic and elaidic acids and designate each as *E* or *Z* isomers.

Q2.4 Allocate the symbol *E* or *Z* to each of the following:

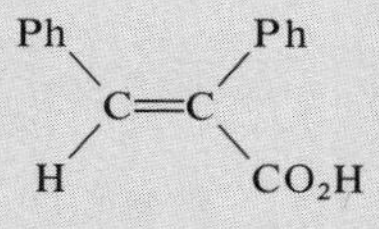

Ph, Ph / C=C / H, CO_2H **(14)**

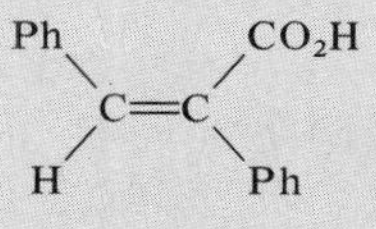

Ph, CO_2H / C=C / H, Ph **(15)**

(Ph = phenyl = C_6H_5)

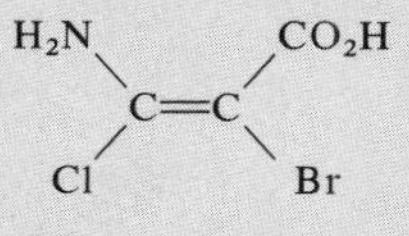

H_2N, CO_2H / C=C / Cl, Br **(16)**

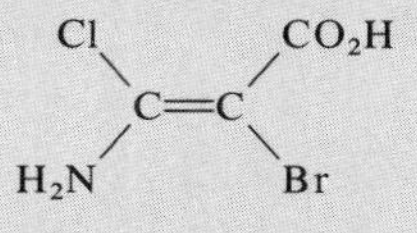

Cl, CO_2H / C=C / H_2N, Br **(17)**

2.5 Is *cis-trans*-isomerism confined to C=C compounds?

So far we have confined our study of *cis-trans*-isomerism to alkenes but one or both carbon atoms can be replaced by another doubly bonded atom. Most commonly this is nitrogen in which one sp^2 orbital contains only a lone pair. For example, oximes and azo compounds can exist in stereoisomeric forms:

isomeric aldoximes: CH_3(H)C=N–OH and CH_3(H)C=N–OH

isomeric ketoximes: C_2H_5(Ph)C=N–OH and C_2H_5(Ph)C=N–OH

isomeric azo compounds: Ph–N=N–Ph and Ph–N=N–Ph

In the past special prefixes have been used to designate these isomers but this is now satisfactorily achieved by the use of (E) and (Z).

Q2.5 Designate the above six structures as (E) or (Z).

Q2.6 What is the structure of the only aldehyde and of those ketones whose oximes *cannot* exhibit *cis-trans*-isomerism? (If in difficulty refer back to section 2 of this chapter.)

Q2.7 So far we have designated structures as (E) or (Z). It is also necessary to be able to write a structure from a name. Formulate the (Z) isomers of the following compounds: (i) but-2-enoic acid, (ii) 4,4′-diaminoazobenzene, (iii) 1,2-difluoroethene, (iv) 2-ethyl-3-methylpent-2-enoic acid, (v) 3,3-diphenylpropenoic acid.

2.6 *Cis-trans*-isomerism in cyclic compounds

Cis-trans-isomerism occurs in any molecule in which four appropriate groups are held in one plane by some element of rigidity within the molecule. So far our examples have involved a double bond such as C=C, C=N, or N=N to achieve this, but it also happens in cyclic compounds in which the ring, though having some flexibility (see Chapter 4), nevertheless restricts the freedom of movement of groups attached to it. In the following discussion the cyclohexane system is treated as planar. This is not so (Chapter 4), but its non-planarity does not weaken the argument and it is simpler at this stage to accept this slight deception.

Cyclohexane-1,4-dicarboxylic acid can be written as structure **18** with the understanding that all the carbon atoms carry as many hydrogen atoms as is required for them to have the usual quadrivalency. However, this compound exists in two forms (mp 162° and 300°) of which only the lower melting isomer readily forms an anhydride. When hydrated the anhydride furnishes the original low melting dibasic acid. This is another example of *cis-trans*-isomerism. The cyclohexane ring can be considered as planar and the four groups (H, CO_2H, H, CO_2H) attached to C(1) and C(4) lie in one plane. The two possible structures are shown (**19** and **20**).

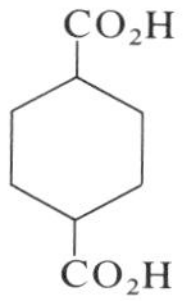

CO_2H CO_2H H H HO_2C CO_2H H CO_2H HO_2C H

(**18**)  (**19**)†  (**20**)†

Q2.8 Assign the symbols (*E*) and (*Z*) to **19** and **20** and indicate which structure represents the acid of lower melting point.

2.7 What are the possibilities of stereoisomerism in compounds with more than one C=C group?

The possibilities of stereoisomerism increase when the molecule contains more than one C=C (or C=N, etc.) group. Each double bond can lead to an independent set of isomers except that in certain symmetrical molecules some isomers may be identical. Thus the unsymmetrical diene octa-2,5-diene exists in four stereoisomeric forms and the symmetrical hepta-2,5-diene in only three.

† These structures contain three types of lines which are used in a standard way when trying to indicate three-dimensional models in a two-dimensional picture. Light lines —— are interpreted as being in the plane of the paper, heavy lines —— project forward from the paper toward the reader, and dashed lines ---- or ···· project backward from the paper away from the reader.

Q2.9 Structure **21** is one stereoisomer of octa-2,5-diene, viz. octa-(E)-2,(E)-5-diene. Complete the other three stereoisomers.

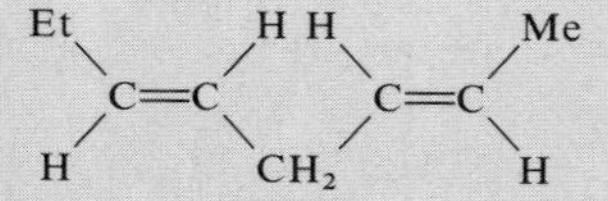

Et, H, H H, CH$_2$, Me, H — C=C, C=C

(21)

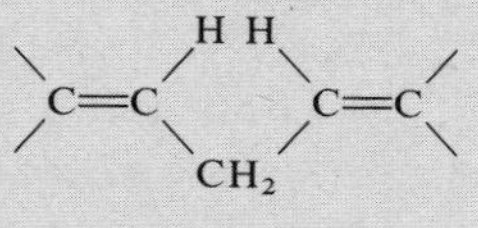

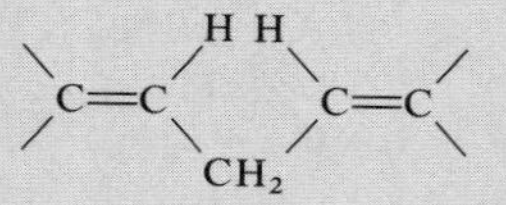

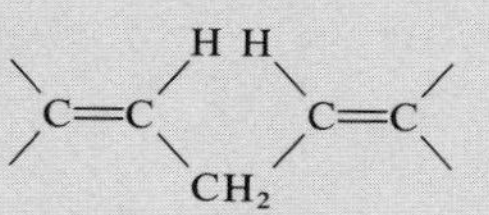

Q2.10 Formulate and name the three stereoisomeric hepta-2,5-dienes.

2.8 How are structural assignments made?

In the course of this chapter we have assigned structures to a few stereoisomeric pairs on the basis of selected evidence. Now we consider briefly the general methods for deciding geometrical configuration. Most procedures depend on observations that have been made with a few compounds of known structure and then generalised. This approach must be used carefully and it may not be wise to rely on a single piece of evidence: it is better to seek supporting evidence. The procedures depend on the physical or chemical properties of the two (or more) isomeric forms (these are discussed in greater detail, Brewster, 1972). In general, *cis*- and *trans*-isomers have different physical properties and similar but not identical chemical properties. Chemically the isomers *may* react at different rates on account of steric factors and they *may* give products with different stereochemistry (Chapter 5).

Using known compounds many physical properties have been correlated with the *cis*- and *trans*-configuration. Sometimes these correlations are restricted to a group of similar compounds, sometimes they apply more widely. Always they should be used with care. For example, *trans*-compounds often have higher melting points than their *cis*-isomers. Spectroscopic procedures – especially infrared and nuclear magnetic resonance – are usually more reliable. Dipole moments are very satisfactory when they can be applied.

Q2.11 The two isomers of 1,2-dichloroethane have dipole moments of 0·00 and 1·89 D. Formulate each of these and relate them with their dipole moment.

NMR spectroscopy may be used to distinguish between (*E*) and (*Z*) isomers of compounds of the type RCH=CHR′ by attention to the coupling constant of the signal for the two olefinic protons. This is higher for (*E*) compounds (12–18 Hz) than for their (*Z*) isomers (7–11 Hz). Infrared and Raman spectroscopy may also be used to distinguish (*E*) and (*Z*) isomers of this type.

Chemical procedures are based on a knowledge of either reaction mechanism or the stereochemistry of a reaction product. This matter is elaborated in Chapter 5. The preparation of oleic acid and elaidic acid by stereospecific hydrogenation procedures is an example of this approach (p. 12) as is the structure-determination of maleic acid based on its readiness to undergo reversible dehydration.

Q2.12 An acid of structure $Cl_3CCH{=}CHCO_2H$ can be reduced to $CH_3CH{=}CHCO_2H$ and hydrolysed to fumaric acid. What is the configuration of each of the acids formulated?

2.9 Can *cis*- and *trans*-isomers be interconverted?

Early in this chapter (p. 9) it was pointed out that *cis-trans*-isomerism is found among alkenes and certain related nitrogen-containing compounds because the double bond provides a considerable barrier to rotation about its axis. The conversion of *cis*- to *trans*-isomers or the reverse process – known as stereomutation – can be effected in two ways: either by a series of chemical reactions by which the alkene is converted to other compounds from which the alkene can be regenerated as the alternative isomer (examples of this are detailed in Chapter 5) or by a *reversible* procedure converting the double bond compound to a single bond compound in which free rotation is possible before regeneration of the alkene as a mixture of the two isomeric forms.

This reversible process may be effected by vigorous heating, by photolysis, or, usually under milder conditions, by a free radical species such as those resulting from iodine, thiols or other sulphur compounds, or nitro-

$$\underset{\textit{cis}\text{-isomer}}{R'(H)C{=}C(R'')H} \underset{}{\overset{+R\cdot}{\rightleftharpoons}} R'(R)(H)C{-}\dot{C}(R'')H \rightleftharpoons R'(R)(H)C{-}\dot{C}(H)R'' \overset{-R\cdot}{\rightleftharpoons} \underset{\textit{trans}\text{-isomer}}{R'(H)C{=}C(H)R''}$$

radical species undergoing inversion through free rotation about the carbon to carbon bond

gen tetroxide. The product is eventually an equilibrium mixture of the two isomers. The *trans*-isomer, being the more stable, usually predominates but pure compounds have to be separated from the reaction mixture.

Answers

A2.1 These four formulations represent only *two* compounds. Structure **2** is the same as **5** and structures **3** and **4** are also identical. If **4** or **5** is turned over it is seen to be the same as **3** or **2** respectively. Structures **2** and **3**, however, are different. In **2** the groups *a* and *c* are on one side of the double bond and the groups *b* and *d* are on the other side. In structure **3** *a* and *d* lie on one side and *b* and *c* on the other.

A2.2 The compound of structure **7** does not show *cis-trans*-isomerism because one olefinic carbon atom carries two identical groups. The remainder will have diastereoisomers represented by **6′**, **8′**, and **9′**.

$$\underset{(\mathbf{6'})}{(a)(b)C{=}C(c)(d)} \qquad \underset{(\mathbf{8'})}{(a)(b)C{=}C(c)(a)} \qquad \underset{(\mathbf{9'})}{(a)(b)C{=}C(b)(a)}$$

A2.3

$$\underset{\text{maleic acid } (Z)}{HO_2C(H)C{=}C(CO_2H)H} \qquad \underset{\text{fumaric acid } (E)}{HO_2C(H)C{=}C(H)CO_2H}$$

$$\underset{\text{oleic acid } (Z)}{CH_3(CH_2)_7(H)C{=}C((CH_2)_7CO_2H)H} \qquad \underset{\text{elaidic acid } (E)}{CH_3(CH_2)_7(H)C{=}C(H)(CH_2)_7CO_2H}$$

In maleic acid, for example, the ligands attached to each alkene carbon atom are put in order of precedence: $CO_2H > H$ for both carbon atoms. The *Z* isomer will be that one in which the groups of higher precedence (CO_2H and CO_2H) are on the *same* side of the double bond: in the *E* isomer the groups of higher precedence are on *opposite* sides of the double bond.

A2.4 **14** and **15** are (*E*) and (*Z*) isomers respectively ($Ph > H$ but $CO_2H > Ph$). This is an example where the simple correspondence between *cis* and (*Z*) does *not* apply. **16** and **17** are (*Z*) and (*E*) isomers respectively ($Cl > N$ and $Br > C$).

A2.5 aldoximes (*Z*) and (*E*) respectively
ketoximes (*E*) and (*Z*) respectively
azo compounds (*Z*) and (*E*) respectively
Remember that the lone pair comes below H in precedence order.

A2.6 Symmetrical carbonyl compounds $R_2C{=}O$ form oximes which cannot exist as *cis-trans*-isomers. The only aldehyde to meet this requirement is formaldehyde ($H_2C{=}O$) but there are many symmetrical ketones of which acetone ($Me_2C{=}O$) is the simplest.

A2.7 (i)

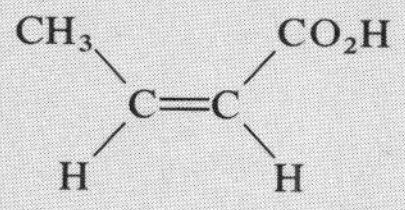

(ii)

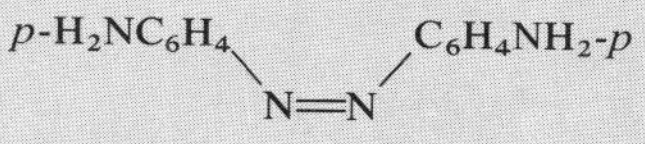

(iii)

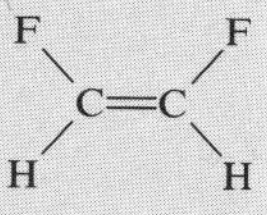

(iv)

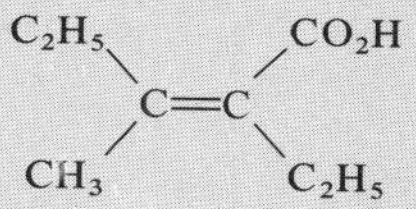

(v) this compound ($Ph_2C{=}CHCO_2H$) does not exist in stereoisomeric forms!
[If you are having difficulty with nomenclature why not try the author's programme "Nomenclature of Aliphatic Compounds" (*Programmes in Organic Chemistry*, Volume 1), 1966, English Universities Press?]

A2.8 **19** and **20** and (*Z*) and (*E*) isomers respectively. By analogy with maleic acid it is expected that anhydride formation will occur most readily when the two CO_2H groups are closest together, i.e. the (*Z*) isomer (**19**).

A2.9

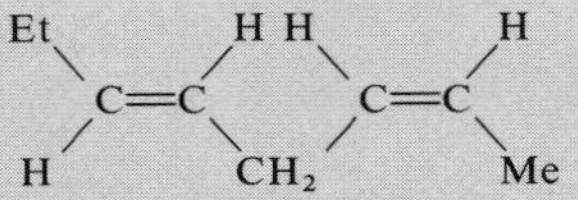

```
Et         H H        H
  \       /   \      /
   C=C          C=C
  /       \   /      \
H          CH2        Me
```
(*Z*)-2, (*E*)-5

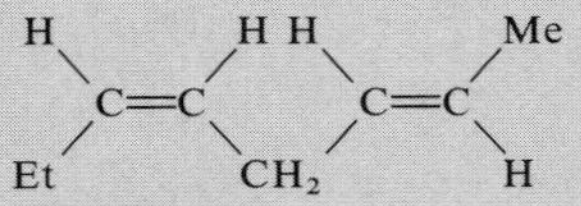

```
H          H H        Me
  \       /   \      /
   C=C          C=C
  /       \   /      \
Et         CH2        H
```
(*E*)-2, (*Z*)-5

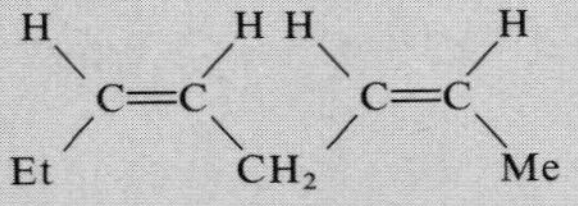

```
H          H H        H
  \       /   \      /
   C=C          C=C
  /       \   /      \
Et         CH2        Me
```
(*Z*)-2, (*Z*)-5

A2.10

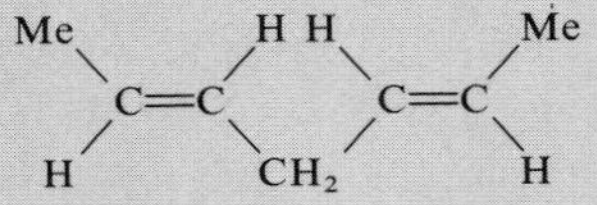

```
Me         H H        Me
  \       /   \      /
   C=C          C=C
  /       \   /      \
H          CH2        H
```
(*E*)-2, (*E*)-5

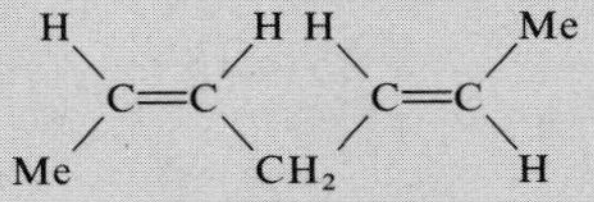

```
H          H H        Me
  \       /   \      /
   C=C          C=C
  /       \   /      \
Me         CH2        H
```
(*E*)-2, (*Z*)-5 or (*Z*)-2, (*E*)-5

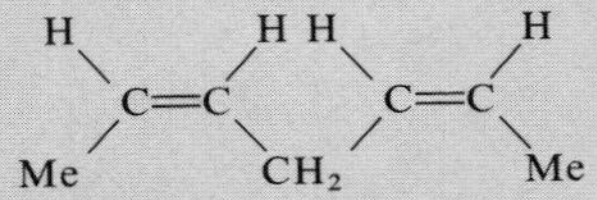

```
H          H H        H
  \       /   \      /
   C=C          C=C
  /       \   /      \
Me         CH2        Me
```
(*Z*)-2, (*Z*)-5

A2.11

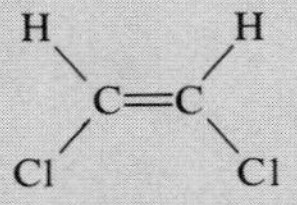

```
H      H
  \   /
   C=C
  /   \
Cl     Cl
```
(*Z*) isomer (1·89 D)

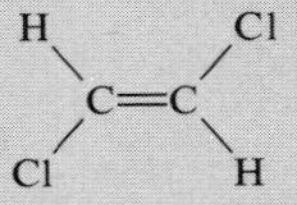

```
H      Cl
  \   /
   C=C
  /   \
Cl     H
```
(*E*) isomer (0·00 D)

The dipole in these molecules is the vector sum of that resulting from the two C→Cl bonds. In the (*E*) isomer these cancel so that it has a zero dipole moment.

A2.12 If it is assumed that these changes ($CCl_3 \rightarrow CH_3$ and $CCl_3 \rightarrow CO_2H$) occur without any change in configuration of the alkene then all the compounds will have the same configuration as fumaric acid, i.e. *trans* or (*E*).

3

Enantiomerism

3.1 Historical introduction

The discovery of the polarisation of light by Malus in 1808 was followed by the realisation by Arago (1811) and by Biot (1813) that certain materials deviated the plane of polarisation. Many of these substances were crystalline inorganic compounds such as quartz and sodium chlorate which displayed this property only in the crystalline form. Others, however, were organic compounds, such as oil of turpentine or cane-sugar or camphor which were active in the solid, liquid (whether molten or in solution), and gaseous state. Such materials were described as optically active and it was noted that whilst some rotated plane polarised light in a clockwise fashion others shifted it in an anticlockwise manner.

The crystallographer Haüy had shown that quartz crystals were of two kinds which also differed in their effect on polarised light. One type of crystal produced a clockwise deviation and the other type of crystal an anticlockwise deviation. Up to 1848, however, no organic compound was known to exist in two forms corresponding to the two types of quartz. It is true that two forms of tartaric acid† were known: one form, derived

† Tartaric acid $HO_2CCH(OH)CH(OH)CO_2H$.

from the potassium hydrogen tartrate which crystallised during fermentation processes, showed a clockwise rotation whereas the so-called racemic acid,† obtained in later stages of the crystallisation process, was structurally identical with tartaric acid but had no influence on polarised light. It was then that the young Pasteur (1822–1895) made an important and exciting discovery: he found that crystals of optically inactive sodium ammonium racemate were of two mirror image crystallographic forms which he could separate manually, that these two forms caused an equal and opposite rotation in solutions of equal concentration, and that rotation was no longer observed when the two types of crystals were mixed in equal amounts. This was the first description of a pair of isomers differing only in ability to rotate plane polarised light in opposite directions.

Many refused to believe Pasteur's results and M. Biot, a veteran physicist and the greatest living authority on polarised light, was delegated by the French Academy of Sciences to examine Pasteur's statements. Fortunately we have Pasteur's record of what happened (Read and Gunstone, 1958):

He [M. Biot] sent for me to repeat before his eyes the several experiments. He gave me racemic acid which he had himself previously examined and found to be quite inactive to polarised light. I prepared from it in his presence the sodium ammonium double salt, for which he also desired himself to provide the soda and ammonia. The liquid was set aside for slow evaporation in one of the rooms of his own laboratory, and when 30 to 40 g of crystals had separated he again summoned me to the College de France, so that I might collect the dextro- and laevo-rotatory crystals before his eyes, and separate them according to their crystallographic character, asking me to repeat the statement that the crystals which I would place on his right hand would cause deviation to the right, and the others to the left. This done, he said that he himself would do the rest. He prepared the carefully weighed solutions, and he again called me into his laboratory. He first put the more interesting solution, which was to cause rotation to the left, into the apparatus. Without making a reading, but already at the first sight of the colour-tints presented by the two halves of the field in the Soleil saccharimeter he recognised that there was a strong laevo-rotation. Then the illustrious old man, who was visibly moved, seized me by the hand, and exclaimed: "My dear child, I have so loved the sciences throughout my life that this makes my heart leap with joy!"

---

† The word racemic (from *racemus*, bunch of grapes) is now used to describe a 1:1 mixture of enantiomers which does not show optical activity because the activity of one form is cancelled by that of its enantiomer.

3.2 Some definitions

After that historical introduction we shall examine this phenomenon at today's level of understanding and use modern nomenclature, but first it will be useful to define some terms which will be used in the ensuing discussion.

(i) Clockwise rotation of polarised light is designated as dextrorotatory and usually abbreviated to *d* or (+), anticlockwise rotation is indicated by the term laevorotatory, *l*, or (−). The symbols D and L do not show the *direction* of rotation but a configurational relationship to some standard substance (usually glyceraldehyde) so that some compounds are D(+) and others are D(−).

(ii) The degree of rotation depends not only on the material being investigated but on the temperature, the solvent, the concentration of the solution, the length of the polarimeter tube, and on the wavelength of the light employed. The **specific rotation** $[\alpha]$ for a stated temperature, wavelength, and solvent is equivalent to the rotation produced by a column 10 cm in length containing 1 g per ml of solution. Thus the symbol $[\alpha]_D^{20^\circ} = +66{\cdot}5$ for an aqueous solution of cane sugar indicates a specific rotation of $+66{\cdot}5^\circ$ for a measurement at 20° using the sodium D-line (5893A). The **molecular rotation** $[M]_D$ is given by

$$[M]_D = [\alpha]\,\frac{M}{100}$$

(iii) Compounds which rotate plane polarised light are described as **optically active**. These always exist in pairs [the (+)-form and the (−)-form] and are said to be **enantiomers** or to be **enantiomeric**. The word **enantiomerism** is now used to describe this type of stereochemistry. Traditionally this phenomenon has been discussed in terms of **asymmetry** and older text-books refer to asymmetric centres and asymmetric compounds. This term has now been replaced by the word **chiral** (see p. 25) and the older term will hardly be employed in this book. The molecule of a compound which is not optically active is **achiral**.

(iv) The reader is reminded (p. 4) that stereoisomers which are not enantiomeric are **diastereoisomeric** and that this latter term is employed both with *cis-trans*-isomers and with isomers which, though optically active, are not enantiomeric.

(v) An equal mixture of two enantiomers produces a **racemic mixture** or **racemate**. This is no longer optically active but racemic mixtures can be separated into their component enantiomers by a process described as **resolution**.

(vi) Definitions of the terms ***meso-*** (p. 39), **pseudoasymmetry** (p. 40), **prochiral** (p. 40), **enantiotopic** (p. 41) and **diastereotopic** (p. 41) are given on the pages indicated.

3.3 Polarimetry

Light can be considered as a wave phenomenon in which vibration occurs at right angles to the direction in which light travels and ordinary light vibrates in all of the infinite number of planes passing through the line of propagation. In plane-polarised light vibration is confined to one of these planes, a result achieved by passing through a polariser consisting of Polaroid or of suitably cut pieces of calcite (Nicol prism).

Some crystals and some other compounds in the solid, liquid, or gaseous state rotate plane-polarised light in a clockwise or anticlockwise manner, i.e. the light emerges from such materials vibrating in a different plane.

Two pieces of Polaroid or two Nicol prisms will only transmit plane-polarised light when suitably aligned. The amount of light transmitted decreases to zero as one prism is rotated through 90° and then increases to

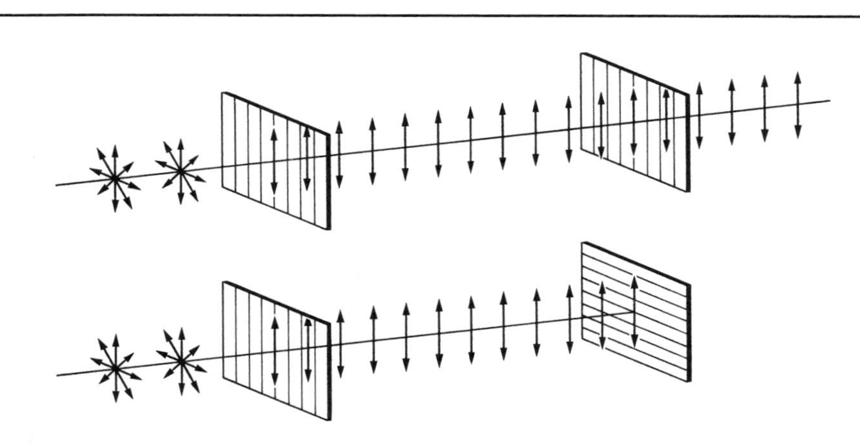

Fig. 2 When a ray of light passes through a Nicol prism or a sheet of Polaroid the light becomes linearly polarised, i.e. it possesses a fixed, well-defined plane of vibration. A second polariser parallel to the first allows the light to pass through it but if the second polariser is perpendicular to the first the light is completely extinguished.

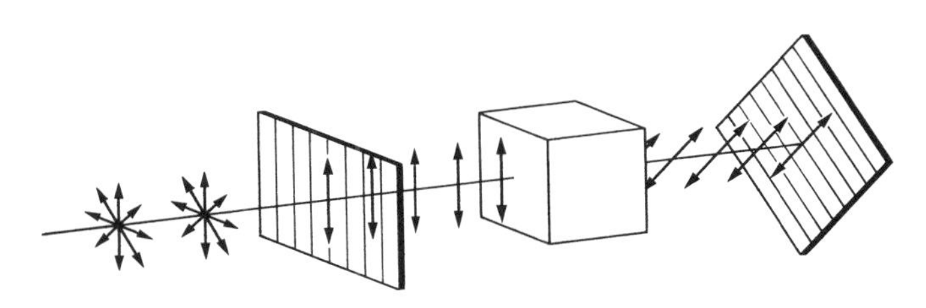

Fig. 3 If a ray of polarised light passes through an optically active substance, the plane of polarisation is rotated through a definite angle. The value of the angle of rotation is measured using a second polariser. The position at which the ray is extinguished is changed with respect to the value obtained in the absence of the active compound. With modern photoelectric polarimeters the angles can be measured with a precision of one ten-thousandth of a degree. The value of the optical activity (or rotatory power) is expressed as specific activity $[\alpha]=\alpha/d.l$, where d is the density of the compound (if it is a pure liquid) or its concentration (if it is a solution) expressed in g/cm^3 and l is the length of the polarimeter tube in decimetres. The rotatory power varies with the temperature and with the wavelength of the light used; thus monochromatic light is used, generally the sodium D line. More accurately, optical activity may be measured at different wavelengths and a graph showing the variation of α as a function of the wavelength λ (an optical rotatory dispersion curve) can then be drawn.

the original value during rotation through another 90°. The extent of rotation produced by optically active material can be determined by discovering to what degree one prism must be rotated to restore the position of maximum or minimum passage of plane-polarised light. These are the essential principles of a polarimeter used to study optical activity.

3.4 Chirality, handedness

We now turn our attention to the question of why, for example, compounds like lactic acid (**1**) or alanine (**2**) each exist in two forms which

$CH_3CH(OH)CO_2H$ (**1**)

$CH_3CH(NH_2)CO_2H$ (**2**)

differ only in that they rotate polarised light in opposite directions and to an equal extent. We have also to face the issue raised in Chapter 1: if there are two forms of lactic acid then there must be two structures to express that difference.

Two compounds which are identical in all respects except that they exert an opposite (but equal) effect on plane-polarised light are said to be **enantiomers** and the whole phenomenon is described as **enantiomerism**. Organic compounds exhibit this property only when their molecules are **chiral**. This term will be explained shortly but it should be noted that enantiomerism occurs only when a molecule contains at least a centre of chirality or an axis of chirality or a plane of chirality. Much of the following discussion will be about molecules having one or more centres of chirality but reference is made later to enantiomers having an axis of chirality (p. 45) and a plane of chirality (p. 47).

To quote the paper of Cahn, Ingold, and Prelog (1966):

The necessary and sufficient condition (of enantiomerism) is that reflexion in a plane converts the model into a non-identical one, that is, one which cannot be superimposed on the original by translations and rotations only. The model has two non-identical forms, inter-related by a reflexion, that is, two enantiomeric forms: it has the topological property of handedness.

To quote another statement (Anon., 1970):

The property of nonidentity of an object with its mirror image is termed chirality. An object, such as a molecule in a given configuration or conformation, is termed chiral when it is not identical with its mirror image; it is termed achiral when it is identical with its mirror image.

More simply: **a molecule is chiral if it cannot be superimposed on its mirror image** or (in more mathematical terms) if it does not possess an alternating axis of symmetry.

The word **chiral** (Greek *cheir* hand, pronounced kīral) implies the property of "handedness", i.e. the object and mirror-image relationship of a left and right hand. This is an important concept which must be clearly grasped. My two hands are similar but *not identical*. My left hand is (more or less) the mirror image of my right hand and *vice versa* but the one cannot be superimposed on the other. The same is true of my feet and also of my gloves and shoes. A right-hand glove fits my right hand but not my left and we learn early in life that it is desirable to have our shoes correctly placed on our feet. This illustrates an important principle that one chiral object (the hand or the foot) can distinguish another chiral object (the glove or the shoe) with which it is designed to interact. This is also true at the molecular level for a drug or an enzyme.

To come back to simple organic chemistry, a molecule having a carbon atom attached to four different ligands is chiral and though, in more exact terms, we should speak of a chiral environment we shall, with less exactitude, refer to the chiral carbon atom and the chiral centre. In the following sections we shall learn to recognise a chiral centre, to determine the number of possible stereoisomers of molecules having more than one chiral centre, to represent these stereoisomers in various ways, to designate them by the terms (*R*) and (*S*), and to distinguish the relative properties of enantiomers, diastereoisomers, and racemic mixtures. Finally we shall discuss features other than chiral centres which may produce chiral molecules.

First let us try to satisfy ourselves that when a carbon atom is attached to four different ligands (i.e. is chiral) the whole structure can exist in two forms.

Q3.1† Using any atomic models available to you attach four different atoms or groups to a carbon atom. Set this model on one side and make another model with the same atoms or groups attached to a carbon atom. Repeat this until you have about six structures. Now examine them carefully to see how many different (non-superimposable) structures you have obtained. Twist and turn the whole structure as much as you like but do not break and re-make any bonds (consult **A3.1** for comment).

Q3.2† In the experiment just completed you examined six structures which could be considered as *Cabcd*. Each had four different ligands attached to carbon. In each structure replace ligand *d* by a second ligand *a* and satisfy yourself that it is not possible to produce two different structures of the type *Caabc*.

3.5 Molecules with one chiral centre

A carbon atom is chiral when it has four different ligands attached to it and the whole molecule then exists in two enantiomeric forms; conversely, if a compound exists in two enantiomeric forms then it must contain at least one chiral centre (unless it has an axis or plane of chirality).

† This exercise can be carried out more quickly if several students each prepare a single structure and then compare them.

Q3.3 Lactic acid (2-hydroxypropanoic) and alanine (2-aminopropanoic acid) each exist in enantiomeric forms. Formulate these structures and indicate in each structure which carbon atom is chiral and what ligands are attached to it.

Notice that in designating the ligand attached to a reference atom we are concerned with the whole group and not just the single atom through which the attachment is made. In the examples we have just considered CH_3 and CO_2H are different ligands even though both are linked to the chiral centre through a carbon atom.

Q3.4 The following structures represent molecules which may have two, one, or no chiral centres. Indicate which carbon atoms (if any) are chiral.

(i) $CH_3CHBrCO_2H$
(ii) $HO_2CCHClCO_2H$
(iii) $CH_3(CH_2)_7CH(OH)(CH_2)_7CO_2H$
(iv) $EtCH_2COOCH(Me)Et$
(v) $BrCH_2CHDCH_2Cl$
(vi) EtCH=CHCH(Me)CH=CHEt (*E*) (*Z*)
(vii) OH Cl

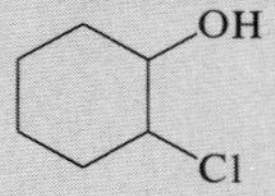

3.6 The properties of enantiomers and racemic mixtures

Compounds whose molecules contain one chiral centre exist in two enantiomeric forms. They have identical physical properties apart from their differing influence on polarised light which they rotate in opposite directions but to an equal extent. They may also smell differently: (+)-limonene smells of oranges and (−)-limonene of lemons, whilst (−)-carvone isolated from spearmint oil and (+)-carvone from caraway seed oil also have different odours. Enantiomers are distinguished by prefixing their name with the symbol (+)- or (−)- as appropriate. The alternative symbols *d*- and *l*- are now hardly ever used. The problem of how they are to be represented on paper will be taken up later. The two enantiomers also have identical chemical properties *except in the presence of chiral reagents* (i.e. in reactions with other optically active compounds which may be simple chemicals or complex substances like the enzymes). A 1:1 mixture of two enantiomers will no longer influence polarised light since the effect of one enantiomer is cancelled by the other. This mixture is called a **racemate** (see p. 23) and may be prefixed by the symbol (±)-. It has the same chemical properties in the liquid state but *may* have different physical properties in the solid state. Lactic acid, for example, melts at 53° in either of its enantiomeric forms but at 17° in its racemic form. Check your understanding of this paragraph in the next question.

Q3.5 The compound A exists in two enantiomeric forms (+)-A and (−)-A and in a racemic form (±)-A. From the information provided complete the following table as far as possible. If a value cannot be predicted from the data then insert a question mark.

	(+)-A	(−)-A	(±)-A
specific rotation	+25°		
melting point		44°	
refractive index (50°)			1·268

Although enantiomeric compounds usually have the same nmr spectra they may show distinctive signals in a chiral environment which causes them to behave as diastereoisomers. This has been achieved in three ways. (i) Reaction of the enantiomers with a single enantiomeric reagent (such as PhCH(OMe)COCl) gives diastereoisomeric derivatives which show different chemical shifts though these differences are often rather small. (ii) In the presence of a chiral solvent (such as 2,2,2-trifluoro-1-phenylethanol or α-(1-naphthyl)ethylamine) two enantiomers have different chemical shifts because of differing solute-solvent interaction. (iii) Enantiomers show different signals in the presence of chiral shift reagents such as transition metal compounds derived from camphor.

3.7 Fischer projections

Before we finish our study of molecules with one chiral centre we consider one more important example – glyceraldehyde – and apply ourselves to the problem of how to represent its two enantiomers.

Glyceraldehyde has the structure **3** and its systematic name is 2,3-

$HOCH_2CH(OH)CHO$
(**3**)

dihydroxypropanal. The aldehyde group is the principal functional group which is therefore designated as a suffix and assigned the lowest possible number, and since this is one it need not be cited. Other equally adequate ways of representing this molecule are given below (**4**–**7**).

$OHCCH(OH)CH_2OH$ (**4**)

CHO–CHOH–CH_2OH (**5**)

CH_2OH–CHOH–CHO (**6**)

OH, O, HO (**7**)

The central carbon atom is chiral and therefore glyceraldehyde will exist in two enantiomeric forms. Can we write two structures to represent the two enantiomers and if we achieve this which structure will represent (+)-glyceraldehyde and which will represent (−)-glyceraldehyde?

Since the two molecules differ only in three dimensional terms then we can represent that difference only by using three-dimensional models or by devising systems for representing three dimensional structures in two dimensional drawings. One way of doing this is shown in structure **8** and **9** (see footnote on p. 15 for an explanation of the significance of the different lines).

CHO, H, $HOCH_2$, OH (**8**)

CHO, H, HO, CH_2OH (**9**)

CH_2OH, OHC, H, OH (**10**)

Make models of structures **8** and **9** to satisfy yourself that these two structures are not superimposable and are therefore different from each other: one structure is the reflexion of the other. These structures could have been drawn in many other ways but any other representation must be identical with **8** or **9**.

Q3.6 Is structure **10** the same as **8** or **9**?

The CHO group (C(1) according to systematic nomenclature) was placed at the top of structures **8** and **9** and the CH_2OH group (C(3) in systematic nomenclature) at the bottom for reasons that will become significant later. The two forms of glyceraldehyde can also be written as **11** and **13** in which the chiral carbon atom lies in the plane of the paper, the

CHO
H—C—OH
CH$_2$OH
(**11**)

CHO
H—┼—OH
CH$_2$OH
(**12**)

CHO
HO—C—H
CH$_2$OH
(**13**)

CHO
HO—┼—H
CH$_2$OH
(**14**)

groups above and below project backward and the groups to the right and left project forward. Rosanoff suggested that these could be simplified and written as **12** and **14** and this idea was extended by Emil Fischer to other carbohydrates. This is the basis of the Fischer projections (such as **12** and **14**) which are commonly employed for representing three-dimensional structures.

If Fischer projections are to be used certain rules must be obeyed and certain conventions clearly understood, otherwise the structure will be misinterpreted.

(i) The structure is written in a vertical rather than a horizontal form and the carbon atom bearing the lower number (in standard nomenclature terms) is normally written uppermost. In glyceraldehyde CHO is placed at the top of the Fischer projection because this is C(1) according to Nomenclature Rules.

(ii) In a Fischer projection the chiral atom under consideration lies in the plane of the paper and although all bonds are represented by plain lines it is understood that groups vertically linked to the chiral atom project below the paper and that groups horizontally linked to the chiral atom project above the paper. This must be clearly understood when visualising Fischer projections.

(iii) For purposes of comparison a Fischer projection may be rotated through 180° in the plane of the paper *but no other manipulation* is permitted.

Q3.7 Write the two Fischer projections for glyceraldehyde and then decide which one is equivalent to structure **A**.

H
OHC—C—CH$_2$OH
OH
(**A**)

Q3.8 Write Fischer projections for the enantiomeric forms of (i) $CH_3CH(Br)CO_2Me$ and (ii) $CH_3(CH_2)_7CH(OH)(CH_2)_7CO_2H$.

3.8 Which structure refers to which enantiomer? The (*R*) and (*S*) system

We can now write structures such as **8** and **9** or Fischer projections **12** and **14** for compounds containing a single chiral centre but we still do not know which enantiomer has which structure. It would also be convenient to have a simple way of describing these two structures without having to draw them in full each time.

It is now possible to determine absolute configuration by means of crystallographic procedures (Bijvoet *et al.*, 1951) which are, however, only applicable in special circumstances. In the case of the two glyceraldehyde structures Rosanoff made a guess. He *assumed* that structure **12** with the OH group to the right represented (+)-glyceraldehyde and that structure

CHO	CHO
H—┼—OH	HO—┼—H
CH_2OH	CH_2OH
(+)-glyceraldehyde	(−)-glyceraldehyde
D-glyceraldehyde	L-glyceraldehyde

14 represented (−)-glyceraldehyde. At a later date these were also designated D-glyceraldehyde and L-glyceraldehyde respectively and all compounds which could be considered to be derivable from D-(+)-glyceraldehyde were designated D, independently of whether their sign of rotation was (+) or (−), and their enantiomers were designated L. This was a useful procedure, though it led to confusion when an enantiomer could be related to (+)-glyceraldehyde by one procedure and to (−)-glyceraldehyde by another. D and L have now been largely replaced by the (*R*) and (*S*) symbols proposed by Cahn and Ingold and based on their Sequence Rules.

According to their proposal any chiral centre can be designated (*R*) or (*S*) by the following procedure:

(i) List the four ligands attached to the chiral centre in decreasing order of precedence.

(ii) View the structure from the side opposite the ligand of lowest precedence and consider the spatial arrangement of the three remaining ligands in order of decreasing precedence.

(iii) This order must be clockwise or anticlockwise and is designated (*R*) (Latin, *rectus*, right) or (*S*) (Latin, *sinister*, left) respectively.

Q3.9 Apply these procedures to the two forms of glyceraldehyde and designate them as (*R*) or (*S*) as appropriate.

Q3.10 Designate structures (**15**) to (**17**) as (*R*) or (*S*). (Try first without a model and then check your answer with a model.)

CH_2OMe
H—+—Br
CH_3
(**15**)

CO_2Me
HO—+—H
CH_2Br
(**16**)

CH_2NH_2
Cl—+—Br
CH_2NHMe
(**17**)

3.9 Compounds with two different chiral centres

The simplest C_3 sugar – glyceraldehyde – was a useful example for discussing several factors about the stereochemistry of compounds with one chiral centre and the next members of this series – the C_4 sugars – provide a convenient starting point for compounds with two different chiral centres.

A molecule with one chiral centre exists in two enantiomeric forms which together form a racemic mixture. In general, molecules with *n* different chiral centres exist in 2^n enantiomeric forms which can be combined in appropriate pairs to form half that number (2^{n-1}) of racemic mixtures. The C_4 sugars therefore, with two different chiral centres, should exist in four enantiomeric forms and as two racemic mixtures. These are known as threose and erythrose and have the Fischer projections **18**–**21**.

CHO
H—+—OH
H—+—OH
CH_2OH
(**18**)
erythrose

CHO
HO—+—H
HO—+—H
CH_2OH
(**19**)
erythrose

CHO
H—+—OH
HO—+—H
CH_2OH
(**20**)
threose

CHO
HO—+—H
H—+—OH
CH_2OH
(**21**)
threose

Erythrose exists in two enantiomeric forms (**18** and **19**) as does threose (**20** and **21**). The compounds represented by **18** and (say) **20** are diastereoisomers. They are stereoisomers which are not enantiomeric.

In extension of what was said earlier about properties (p. 28) the enantiomers **18** and **19** differ from one another in only one physical property – the direction in which they rotate plane polarised light – and in no chemical property save those with chiral reagents. The same statement is true of **20** and **21** but the properties of **18** and (say) **20** will be similar but not identical. The two racemic C_4 sugars will also have similar but not identical properties.

Q3.11 Write Fischer projections for all the enantiomers of 2-bromo-3-chlorobutane ($CH_3CHBrCHClCH_3$) and pair them off in racemates.

Q3.12 Assuming that A and B represent different forms of a molecule with two different chiral centres, complete the following table of properties as far as possible. Insert a question mark where the property cannot be predicted.

	(+)-A	(−)-A	(±)-A	(+)-B	(−)-B	(±)-B
specific rotation		−14°				
melting point			56°			63°

Q3.13 To name the compounds represented by structures **18**–**21** each chiral centre must be designated (*R*) or (*S*). Each will then be of a form such as (*R*)-2,(*S*)-3,4-trihydroxybutanal. Name each structure in this way.

3.10 The prefixes *threo*- and *erythro*-

Other molecules with two different chiral centres, not necessarily on adjacent carbon atoms, are frequently designated by the prefix *threo*- or *erythro*-. These are to be interpreted as meaning that two identical (or similar groups are on the same side (*erythro*) or on opposite sides (*threo*) of the carbon chain *when expressed as a Fischer projection.*

It is important to realise that the Fischer projection represents a structure *according to an agreed convention* and does not necessarily indicate the conformation (see Chapter 4) which the molecule will assume in the crystalline state, in the liquid state, or in a solvated form. It does not follow, for example, that two hydroxyl groups on adjacent carbon atoms are closer together in the *erythro*-isomer than in the *threo*-isomer.

Q3.14 Which prefix, *threo*- or *erythro*-, is correctly applied to structures **22–24**?

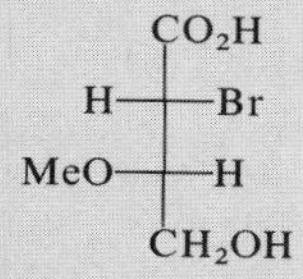

(22): CO_2H at top; H–C–Br; MeO–C–H; CH_2OH at bottom

(23): CO_2Me at top; H–C–OMe; H–C–H; H–C–OMe; CH_2Br at bottom

(24): CO_2H at top; H–C–Br; Br–C–H; Ph at bottom

(22) 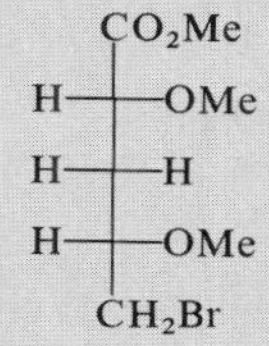 **(23)** 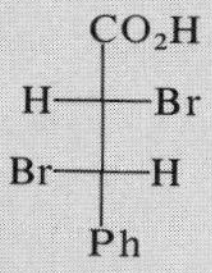 **(24)**

3.11 Stereoregular polymers

Polymerisation of vinyl compounds leads to long-chain polymers with a large number of chiral centres. When there is no pattern in the steric dis-

$$CH_2{=}CHX \longrightarrow \cdots CH_2CH(X)CH_2CH(X)CH_2CH(X)CH_2CH(X)\cdots$$

tribution of the additional group X the polymer is said to be disordered or **atactic**. Stereoregular polymers are described as **tactic** and may be **isotactic** or **syndiotactic** depending on whether the substituent is always on the same side of the carbon backbone in a Fischer projection or whether the substituents are arranged alternately. In reality, the long-chain molecules tend to assume a helical form.

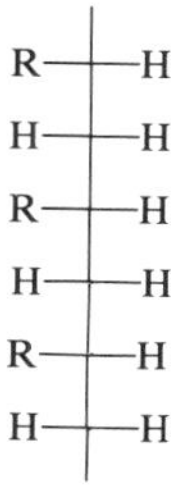

isotactic	syndiotactic
R–H	R–H
H–H	H–H
R–H	H–R
H–H	H–H
R–H	R–H
H–H	H–H

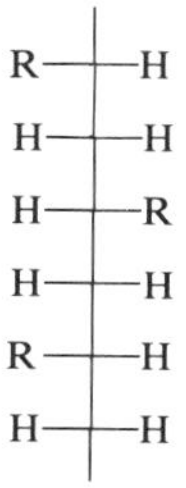

Organometallic catalysts furnish products with a high degree of stereospecificity and polypropylene, polybutene, and polystyrene, exist as isotactic polymers. Polybutadiene and polypropylene prepared under other conditions are syndiotactic. Stereoregular polymers have exceptionally high crystallinity, density, and melting point.

3.12 Sawhorse projections

When it is necessary to emphasise the spatial relationship between ligands attached to two adjacent atoms two other conventions besides the Fischer projection may be used. Each has its merits and it is important to understand all these projections and to be able to interconvert them.

The sawhorse projection makes use of symbols such as **25** and **27** which are more correctly represented as **26** and **28**.

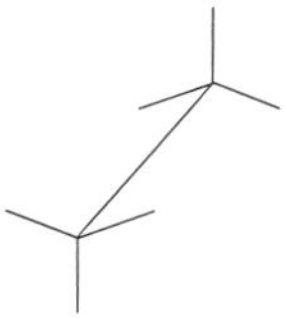

(**25**) 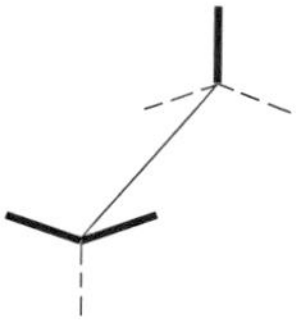 (**26**) 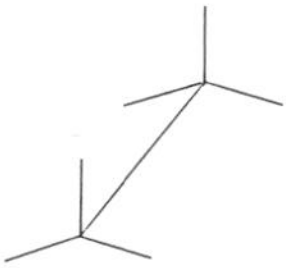 (**27**) 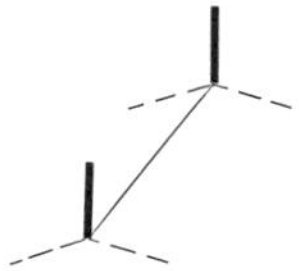 (**28**)

The two atoms under consideration are at the two points where four lines intersect and the remaining ligands are attached to the six free ends. The bond linking the two key atoms is considered to be in the plane of the paper and the remaining lines (bonds) project above (heavy lines) or below (dashed lines) that plane. There is normally free rotation about a single bond and the three groups attached to the front atom may be rotated clockwise or anticlockwise in relation to the three groups attached to the rear atom or *vice versa*.

Let us convert the erythrose enantiomer of Fischer projection **18** to a sawhorse projection. Remembering the disposition of groups in the

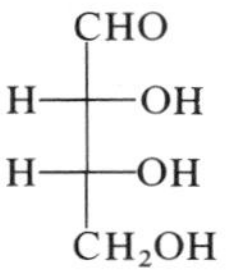

CHO
H—OH
H—OH
CH_2OH
(**18**)

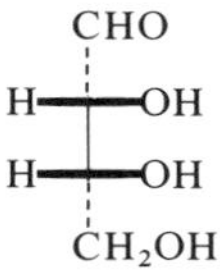

CHO
H—OH
H—OH
CH_2OH
(**29**)

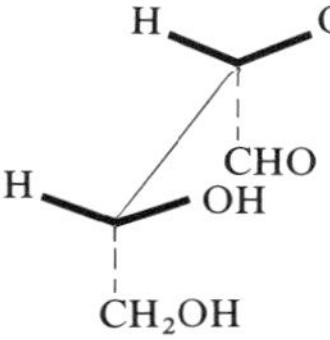

H OH
CHO
H OH
CH_2OH
(**30**)

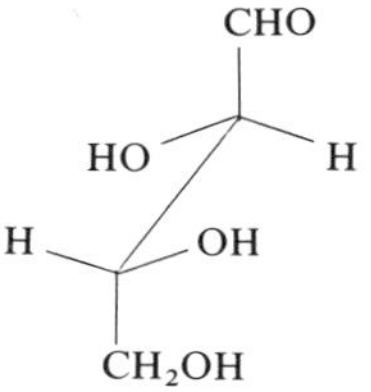

CHO
HO H
H OH
CH_2OH
(**31**)

Fischer projection we first redraft this as **29** and then as the sawhorse projection **30**. Since the carbon chain in acyclic compounds commonly assumes a zigzag conformation, this enantiomer of erythrose can be represented as **31** in which the rear half of the structure has been rotated with respect to the front. Eventually these transformations can be made quickly and in one step but in the beginning it is better to proceed slowly and carefully.

Q3.15 Repeat the process for the enantiomer of erythrose (structure **19**) and show that the structure obtained is the reflexion of structure **31** and that these cannot be superimposed. (This must be so since **18** and **19** represent enantiomers.)

Q3.16 Repeat this process for the threose enantiomer (**20**) and satisfy yourself that whilst the sawhorse projection obtained cannot be superimposed on those from **18** or **19**, neither is it a reflexion of these.

Q3.17 Since this conversion must be done in either direction convert the following structures to Fischer projections.

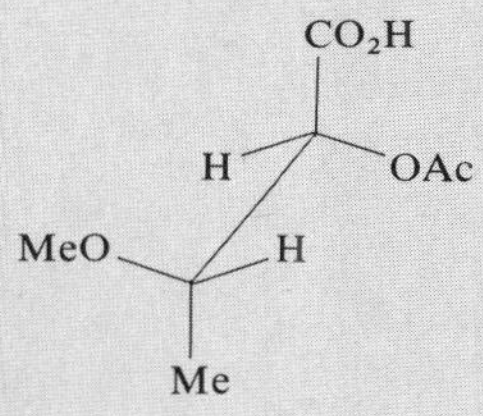

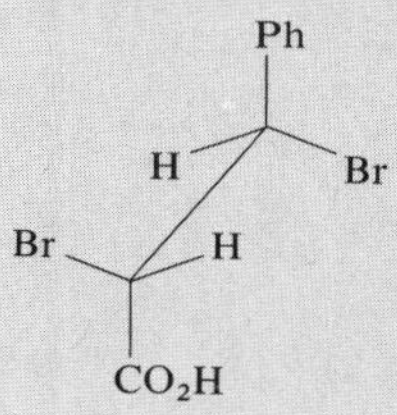

3.13 Newman projections

Like the sawhorse projection, the Newman projection shows the spatial relationship between the ligands attached to two adjacent atoms. Symbols such as **32** or **33** are used. In both of these solid lines are used to indicate

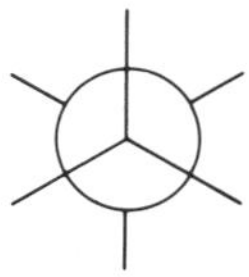

(32) 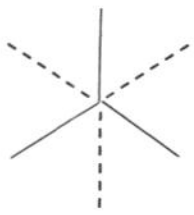 (33) 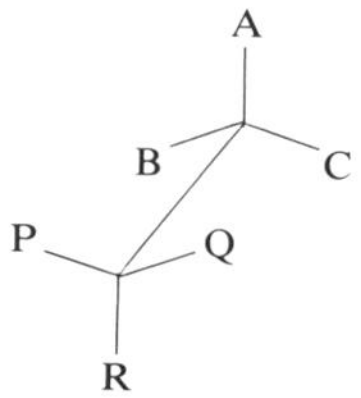 (34) 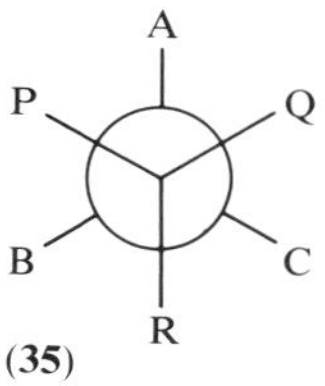 (35)

the ligands attached to the nearer atom and the incomplete lines in structure **32** and the dashed lines in structure **33** are used for ligands attached to the rear atom. Sawhorse projection **34** can therefore be transposed to Newman projection **35**. Since there is free rotation about the single bond joining the two reference atoms projection **35** is but one out of many ways of representing one stereoisomer.

Q3.18 How many different stereoisomers are represented by structures **36** to **40**?

(36) (37) (38)

(39) (40)

Q3.19 Complete structures **41** and **42** as Fischer projections of structures **36** and **37** to satisfy yourself that these are different.

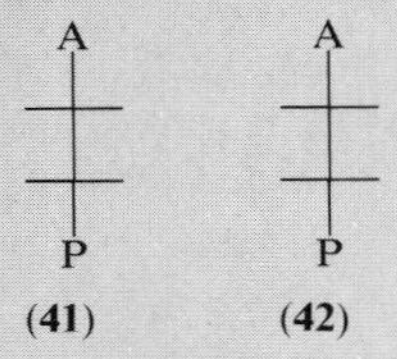

(41) (42)

Q3.20 Write a Fischer projection for one enantiomer of *threo*-2,3-dichloro-3-phenylpropanoic acid ($PhCHClCHClCO_2H$) and complete the following sawhorse and Newman projections for this same enantiomer.

CO_2H

Ph

Ph

CO_2H

3.14 Compounds with two identical chiral centres, *meso* compounds

A new feature arises when a molecule contains two *identical* chiral centres. Consider the compound 2,3-dibromosuccinic acid. This molecule contains two chiral centres but the four ligands attached to one chiral carbon atom (CO_2H, H, Br, and $CHBrCO_2H$) are the same as those attached to the second chiral atom. Following our experience with threose and erythrose we can write four Fischer projections for this compound. Structures

(43)	(44)	(45)	(46)
CO_2H	CO_2H	CO_2H	CO_2H
H—Br	Br—H	H—Br	Br—H
Br—H	H—Br	H—Br	Br—H
CO_2H	CO_2H	CO_2H	CO_2H

43 and **44** are *threo* isomers and **45** and **46** are *erythro*-isomers. The *threo* pair are truly enantiomers, they are non superimposable and one is the reflexion of the other. A closer look at **45** and **46**, however, will show that they are in fact the same. If one structure is rotated through 180° it becomes identical with the other. This stereoisomer contains two chiral centres but they are identical and cancel each other. The molecule as a whole is achiral, and this stereoisomer, represented equally correctly by structure **45** or **46**, is not optically active. How many isomers of 2,3-dibromosuccinic acid are there? This compound can exist in the (+)-*threo* form, the (−)-*threo* form, as a racemic mixture of these two, and as the inactive *erythro* form. This latter is frequently described as the *meso* form and in common with other *meso* compounds *contains a plane of symmetry dividing the molecule into two identical parts such that one is the reflexion of the other.* There is no such plane of symmetry in the molecules of the *threo* compounds.

Q3.21 Another molecule of this type is tartaric acid ($HO_2CCH(OH)CH(OH)CO_2H$). Draw Fischer projections for the stereoisomers of tartaric acid. Which form did Pasteur use in his resolution experiments (p. 22)?

3.15 Pseudoasymmetry

The situation becomes even more complex with a molecule like trihydroxyglutaric acid (2,3,4-trihydroxypentanedioic acid) which exists in four enantiomeric forms (combined appropriately into two racemates) and in two different *meso*-isomers. The molecule contains three chiral centres and we can therefore write eight Fischer projections thus:

(47)	(48)	(49)	(50)
CO_2H	CO_2H	CO_2H	CO_2H
HO—H	H—OH	HO—H	H—OH
HO—H	H—OH	HO—H	H—OH
HO—H	H—OH	H—OH	HO—H
CO_2H	CO_2H	CO_2H	CO_2H

(51)	(52)	(53)	(54)
CO_2H	CO_2H	CO_2H	CO_2H
HO—H	H—OH	H—OH	HO—H
H—OH	HO—H	HO—H	H—OH
H—OH	HO—H	H—OH	HO—H
CO_2H	CO_2H	CO_2H	CO_2H

Q3.22 Which of the above structures (**47**–**54**) represent the four optically active forms of trihydroxyglutaric acid and which represent the two meso forms?

The central carbon atom in this molecule is linked to two different ligands and to an enantiomeric pair of ligands, i.e. it may be written as C(+*a*)(−*a*)*bc* or as C(+*a*)(+*a*)*bc*. Such an atom is said to be pseudo asymmetric when it has the form C(+*a*)(−*a*)*bc*.

3.16 Prochiral centres

An atom linked to two identical ligands and to two other different ligands (*Caabc*) is said to be **prochiral**. If one of the *a* ligands is replaced by another ligand different from any of the others then a new chiral centre has been produced (C*abcd*). Both ethanol and propanoic acid, for example, contain prochiral centres. This is an important feature of many biological reactions

$CH_3\overset{*}{C}H_2OH$ ethanol $CH_3\overset{*}{C}H_2CO_2H$ propanoic acid

since it appears that enzymes (themselves chiral) can sometimes distinguish between the two identical ligands and replace only one of them. The identical ligands (most commonly H) can be designated as pro-(*R*) and pro-(*S*) depending on whether the (*R*) or (*S*) enantiomer results when the ligand is replaced by one of higher precedence without changing the precedence of the remaining ligands.

The two identical ligands attached to a prochiral centre are said to be **enantiotopic** if their separate replacement by a group different from any attached to the prochiral centre gives rise to a pair of enantiomers and to be **diastereotopic** if diastereoisomers are obtained on replacement.

CO_2H / H_S—C—H_R / CH_3 ⟶ CO_2H / D—C—H / CH_3 CO_2H / H—C—D / CH_3 (enantiomers)

enantiotopic ligands at a prochiral centre   (*S*)   (*R*)

CO_2H / H_R—C—H_S / H—C—Br / CH_3 ⟶ CO_2H / D—C—H / H—C—Br / CH_3 CO_2H / H—C—D / H—C—Br / CH_3 (diastereoisomers)

diastereotopic ligands at a prochiral centre

3.17 Can enantiomers be interconverted?

The short answer to this question would be "no" for there is no simple and general procedure for converting a stereoisomer to its enantiomeric form. Nevertheless this answer is not quite complete and there are two additional points to be made.

It is sometimes possible to devise a series of reactions, all of which occur with a high degree of stereospecificity, so that the overall result of the sequence is to change one enantiomer to the other. Some examples of this will be given in Chapter 5.

In addition there are several procedures whereby an enantiomer [(+)-A] can be converted to its racemate [(±)-A] and then separated into its two constituents [(+)-A and (−)-A]. These two processes are described as **racemisation** and **resolution** respectively.

Racemisation is sometimes achieved by vigorous heating or through the catalytic activity of an acid, a base, or some metallic compound or by conversion to an achiral compound and regeneration in the racemic form. As an example of this last procedure an enantiomeric alcohol can be oxidised to an achiral ketone and then reduced to the alcohol in racemic form:

(+)-RCH(OH)R′	⟶	RCOR′	⟶	(±)-RCH(OH)R′
chiral molecule enantiomer		achiral molecule		chiral molecule racemic mixture

The alcohol function must, of course, be part of the chiral centre otherwise it would be destroyed during oxidation.

A familiar type of enantiomeric compound which is readily racemised is one in which the chiral atom is attached both to hydrogen and to a carbonyl group. Such compounds are chiral in the keto form but achiral in the enol form. Since regeneration of the keto form is not stereospecific the interconversion is accompanied by racemisation. Both acids and bases

R(H)C(R′)(COR″)	$\xrightarrow{\text{enolisation}}$	R(R′)C=C(OH)(R″)	$\xrightarrow{\text{ketonisation}}$	R(H)C(R′)(COR″)
(+)-keto form (chiral)		enol form (achiral)		(±)-keto form (chiral)

catalyse the keto-enol equilibrium and therefore promote racemisation. Enantiomers of this type may not, therefore, be optically stable and may racemise – slowly or quickly – when stored.

3.18 Can enantiomers be obtained from racemic mixtures?

Compounds which contain chiral centres are usually produced in the racemic form by chemical synthesis but in an enantiomeric form from natural sources. Occasionally both enantiomers are found naturally in different sources and sometimes the compound is present in racemic form. Enantiomeric compounds produced in nature result from the activity of enantiomeric substrates and/or enantiomeric reagents (extending that term to include the biological catalysts (enzymes) which enable reactions to proceed under the limited range of pH and temperature usually associated with physiological conditions). Most reactions conducted *in vitro*, on the other hand, are between achiral reagents and achiral substrates and even when the product contains one or more chiral centres the product is always a racemic mixture of the possible enantiomers. This is because chemical reagents cannot usually distinguish between the two sides of a double bond or between prochiral atoms.

The preparation of a synthetic enantiomer therefore requires an additional step. Having made the desired compound in racemic form it has to be separated into its enantiomeric forms, a process described as **resolution**. This has been achieved in three ways of which the last is capable of the widest application. All of them were pioneered by Pasteur.

(i) In a few unusual cases enantiomeric molecules form enantiomeric crystals each of which contains only one kind of molecule. These crystals can be distinguished visually and separated manually. It was Pasteur's good fortune that sodium ammonium tartrate is one of the few examples of this: it was his experimental skill and care that enabled him to observe this and to bring his experiments to a successful conclusion (p. 22). It is more common for 1:1 mixtures of enantiomers to crystallise as racemic compounds which cannot be resolved by mechanical means.

(±)-Lactic acid was first resolved as the zinc ammonium salt by Purdie (1892) by a slight modification of Pasteur's procedure in which a supersaturated solution of the racemic salts was seeded with a crystal of one of the enantiomeric forms.

(ii) Enzymes, unlike the usual chemical reagents, are chiral molecules and react differently with enantiomers. Pasteur again was the first to exploit this when, in 1858, he discovered that during the fermentation of ammonium racemate the green mould *Penicillium glaucum* destroyed (metabolised) the (+)-enantiomer of ammonium tartrate and left the (−)-enantiomer practically untouched. More recently it has been demonstrated that the enzyme α-chymotrypsin hydrolyses the acetyl derivatives of (*S*)-α-amino acids but leaves the derivative of the (*R*)-acids untouched.

$$(R,S)\text{-}CH_3CH(NHAc)CO_2H \xrightarrow{\alpha\text{-chymotrypsin}} (S)\text{-}CH_3CH(NH)_2CO_2H + (R)\text{-}CH_3CH(NHAc)CO_2H$$

The disadvantages of this method are that a different enzyme has to be discovered for each substrate (or at least for each class of substrates) and that one enantiomer may be destroyed.

(iii) The third method is based on the fact that, unlike enantiomers which do not differ in their physical properties apart from their influence on polarised light, diastereoisomers display different physical properties. This procedure therefore involves three stages:

(a) conversion of a racemic mixture to two diastereoisomers,
(b) separation of these two diastereoisomers into pure components,
(c) regeneration of the two enantiomers from the separated diastereoisomers.

The conversion of a racemic mixture to a pair of diastereoisomers is effected by reaction of the racemate with some enantiomeric compound (usually a natural compound readily available in enantiomeric form) thus:

$$2(\pm)\text{-A} + 2(-)\text{-B} \longrightarrow (+)\text{-A}(-)\text{-B} + (-)\text{-A}(-)\text{-B}$$

Notice carefully that the products of this reaction are diastereoisomeric and not enantiomeric. The enantiomer of (+)-A(−)-B would be (−)-A-(+)-B. The two diastereoisomers are most often separated by crystallisation but chromatographic and other procedures may also be employed.

Compounds first resolved by this method were racemic acids which form salts with readily available enantiomeric bases such as brucine or cinchonidine or alternatively racemic bases which form diastereoisomeric salts with enantiomeric acids such as tartaric or camphorsulphonic. The process would be summarised as:

2(±)-acid + 2(−)-base ⟶ (+)-acid(−)-base + (−)-acid(−)-base

↓ separation by crystallisation

(+)-acid(−)-base and (−)-acid(−)-base

↓ H^+ ↓ H^+

(+)-acid (−)-acid

Suitable reagents have to be found for other classes of compounds. Thus aldehydes and ketones may be treated with enantiomeric reagents such as menthylhydrazine or menthylsemicarbazide to give diastereoisomeric products. The isocyanate PhCH(Me)NCO has been used for the resolution of chiral amines and alcohols and the latter have often been separated after conversion to acidic compounds such as the half esters of phthalic acid thus:

phthalic anhydride (CO, O, CO) —(±)-ROH→ racemic mixtue of acid phthalates (CO_2R-(±), CO_2H) —(−)-B, e.g. brucine→ diastereoisomeric mixture of brucine salts (CO_2R-(±), CO_2^- (−)-$\overset{+}{B}H$)

—crystallisation→ (CO_2R-(+), CO_2^- (−)-$\overset{+}{B}H$) and (CO_2R-(−), CO_2^- (−)-$\overset{+}{B}H$)

↓ acid ↓ acid

(CO_2R-(+), CO_2H) (CO_2R-(−), CO_2H)

↓ hydrolysis ↓ hydrolysis

(+)-ROH (−)-ROH

In a recently described analytical procedure for determining the amount of (+)- and (−)-alcohols in a mixture these were esterified with (−)-menthyl chloroformate and the diastereoisomeric menthoxycarbonyl esters were separated by GLC. Good separations were effected, though in this case the diastereoisomers were not collected nor were the enantiomers regenerated.

(±)-ROH + (−)menthylchloroformate ⟶ (+)-$ROCO_2$menthyl-(−) and (−)-$ROCO_2$menthyl-(−) (diastereoisomeric menthoxycarbonyl esters)

3.19 Optically active compounds with an axis of chirality

This chapter has been concerned so far only with compounds which are optically active because they contain one or more centres of chirality. We now direct our attention to compounds having an axis or plane of chirality and no centre of chirality. We shall discuss two types of compounds with an axis of chirality: the allenes and certain related compounds on the one hand and the biphenyls on the other.

3.20 Allenes and related compounds

When considering *cis-trans*-isomerism we noted that in the alkene *ab*C=C*ab* the four substituents *abab* lie in one plane and can be arranged in two ways which give rise to *cis*- and *trans*-isomers. As far back as 1875 van't Hoff recognised the possibility of enantiomerism in allenes of the type *ab*C=C=C*ab* but it was 1935 before this was verified experimentally. Since that date natural enantiomeric allenes have also been recognised.

chiral axis ⟶ *ab*C=C=C*ab* ⟵ chiral axis

(55)

*ab*C=C=C*ba*

(56)

(Ph)(αNp)C=C=C(Ph)(αNp)

Ph = phenyl
αNp = α-naphthyl

(57)

HC≡CC≡CCH=C=CHCH=CHCH=CHCH$_2$CO$_2$H

(58)

In an allene the four substituents *abab* attached to the allenic carbon atoms lie in intersecting planes and the two structures **55** and **56** are non-superimposable. The chiral axis in **55** is shown by the two arrows. Typical examples of enantiomeric allenes are the synthetic hydrocarbon **57** and the natural antibiotic mycomycin **58**.†

Since a ring system can produce the same element of rigidity in a molecule one or both of the double bonds of the allene system can be replaced by rings, and compounds such as **59** and **60** also exist in enantiomeric forms because they contain an axis of chirality.

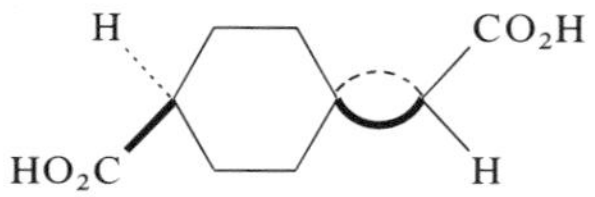

H CO_2H HO_2C H

(59)

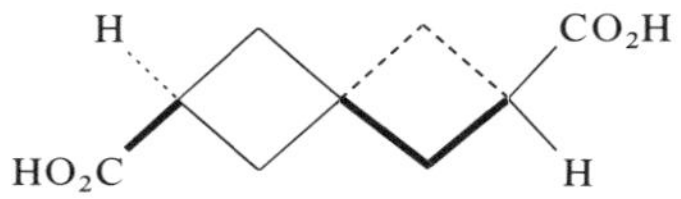

H CO_2H HO_2C H

(60)

3.21 Biphenyls

In the molecule of biphenyl there exists a possibility of rotation about the single bond linking the two phenyl groups so that biphenyl could conceivably exist in a form **(61)** in which the two rings are coplanar, in a form **62** in which the two rings are perpendicular to each other, and in any form between these two extremes.

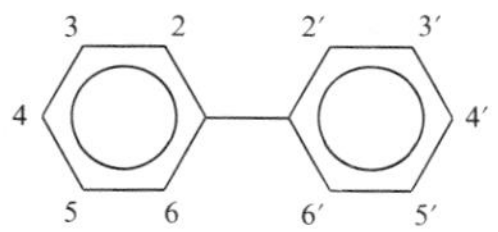

3 2 2′ 3′ 4 4′ 5 6 6′ 5′

(61)

(62)

In the coplanar form the overlap of π orbitals of the two rings allows maximum resonance stabilisation but there is some inhibiting steric interaction of the *o*-hydrogen atoms (2 and 2′, 6 and 6′). In the perpendicular form resonance stabilisation is at a minimum but so also is steric interaction. Electron diffraction data suggest that the two rings are normally inclined at about 40°.

The steric interactions become more significant as the *o*-hydrogen atoms are replaced by bulkier groups and in diphenic acid **(63)**, for example, the angle between the two rings is about 60°.

† This compound, isolated from the soil actinomycete *Norcadia acidophilus*, is of interest in that it contains every possible type of carbon to carbon bonding.

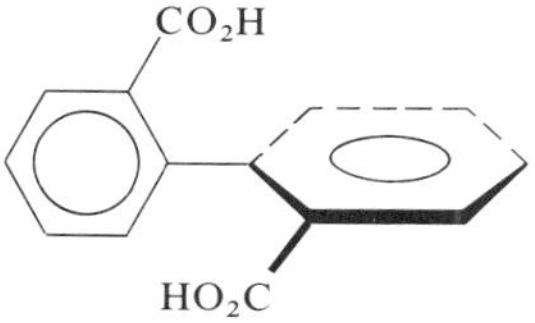

(63) 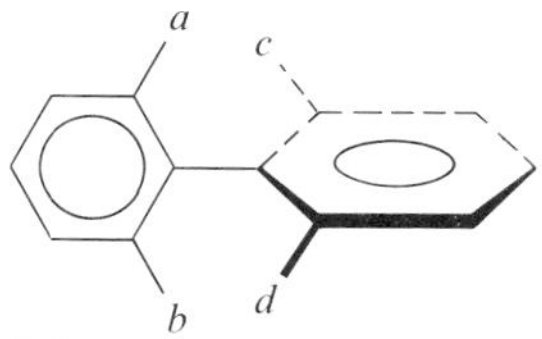 (64)

In a suitably substituted biphenyl compound the non-planar form exists in two enantiomeric structures and since the steric hindrance to rotation depends on the size of these substituents enantiomeric biphenyls vary in their optical stability.

In summary, there are two conditions for enantiomerism in the biphenyl (**64**): (i) The molecule must not contain a plane of symmetry: this requirement is met either if $a \neq b$ and $c \neq d$ or if $a \neq b$ and $c = d$ with a *m*-substituent in this ring. (ii) There must be a sufficiently large barrier between the two enantiomeric forms to preclude rapid interconversion. This is not an all-or-none effect: the rate of interconversion depends on the nature of the *o*-substituents and compounds of varying optical stability are known.

Most enantiomeric biphenyls have three or four substituents in the *o*-positions to provide the necessary optical stability. An unusual example with a single *o*-substituent is the compound with structure **65**.

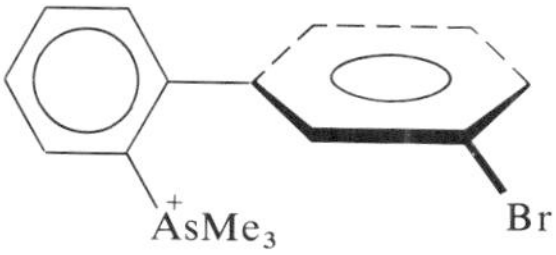

(65)

3.22 Optically active compounds with a plane of chirality

If a molecule has a plane of symmetry which is destroyed by the presence of a suitably placed group then the molecule has a plane of chirality and can exist in enantiomeric forms. In the compound represented by **66**, for example, the alicyclic ring is too small to allow the formation of a planar configuration and the aromatic ring therefore takes up an orientation perpendicular to the alicyclic ring. When R=H the molecule has a plane of symmetry bisecting the aromatic ring but this is destroyed when R is some other substituent such as CO_2H. The same is true of the molecule

(66)

(67)

represented by **67** which exists in enantiomeric forms when R is a substituent other than H.

3.23 Helicenes

An unusual example of chirality is provided by the helicenes. As this trivial name suggests, these molecules assume a helical or spiral arrangement and exist in enantiomeric forms depending on whether they are like a right- or left-handed screw. Of added interest is the fact that these compounds have the highest optical rotation yet recorded.

In the simplest helicene made of six *o*-fused benzene rings planarity is prevented by the presence of carbon atoms which, in a planar arrangement, would occupy the same position and the molecule takes up a helical form. Compounds up to 13-helicene (**68**) are known and have been described as molecular spiral staircases. Enantiomeric helicenes have been obtained by resolution (using an enantiomeric complexing agent), by the selection of enantiomeric crystals, or by homologation of enantiomeric precursors.

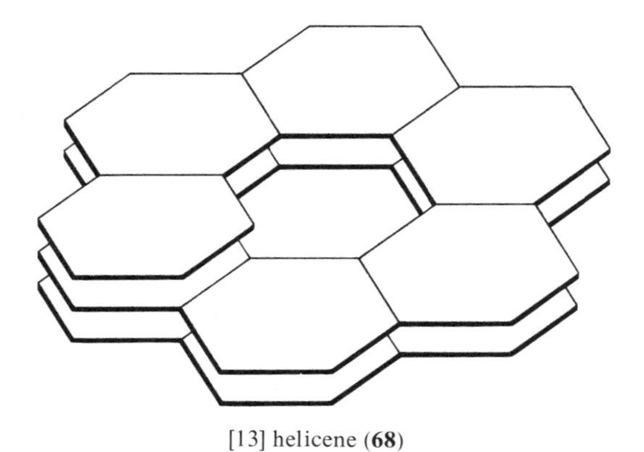

[13] helicene (**68**)

Answers

A3.1 You should have produced two different structures which cannot be superimposed. Notice, however, that one structure is a reflexion of the other. If you were to make another six models – or another six hundred – you should still only have two different structures.

In the exceptional case you may find that all six structures are alike. If this happens break them down and try again. If you have more than two structures then look at them again carefully. You either failed to recognise that some are identical or you did not follow the original instruction to use the same four ligands in each structure.

A3.3 In both structures (**1** and **2** respectively, p. 25) the central atom is chiral and the attached ligands are:

lactic acid:	CO_2H	CH_3	OH	H
alanine:	CO_2H	CH_3	NH_2	H

A3.4 (i) $CH_3\overset{*}{C}HBrCO_2H$ (ii) no chiral centre

(iii) $CH_3(CH_2)_7\overset{*}{C}H(OH)(CH_2)_7CO_2H$ (the two larger ligands are very similar but not identical)

(iv) $EtCH_2COO\overset{*}{C}H(Me)Et$

(v) $BrCH_2\overset{*}{C}HDCH_2Cl$ (H and D are different atoms)

(vi) $\underset{(E)}{EtCH{=}CH}\overset{*}{C}H(Me)\underset{(Z)}{CH{=}CHEt}$ (the difference in configuration of the alkenes is sufficient to make these two unsaturated ligands different)

(vii)

OH

Cl

(this is a cyclohexane derivative not an aromatic compound)

A3.5

	(+)-A	(−)-A	(±)-A
specific rotation	+25°	−25°	0
melting point	44°	44°	?
refractive index (50°)	1·268	1·268	1·268

A3.6 It is identical with **8**. This is not easy to see from the written figures: it is much easier to answer this question if you make a model of **10**.

A3.7

CHO | CHO
H—OH | HO—H
CH_2OH | CH_2OH

A is identical with the Fischer projection on the left. If you have any doubt about this construct a model of **A** and then place it in the correct position for translation to a Fischer projection.

A3.8

CO_2Me | CO_2Me | $(CH_2)_7CO_2H$ | $(CH_2)_7CO_2H$
H—Br | Br—H | H—OH | HO—H
CH_3 | CH_3 | $(CH_2)_7CH_3$ | $(CH_2)_7CH_3$

A3.9 D(+)-glyceraldehyde is (*R*) and its enantiomer (*S*). (If you have difficulty with this refer back to the rules and use a model.)

A3.10 **15** (*R*) **16** (*R*) **17** (*S*)

A3.11

CH_3 | CH_3 | CH_3 | CH_3
H—Br | Br—H | H—Br | Br—H
H—Cl | Cl—H | Cl—H | H—Cl
CH_3 | CH_3 | CH_3 | CH_3

(first pair: and; second pair: and)

A3.12

	(+)-A	(−)-A	(±)-A	(+)-B	(−)-B	(±)-B
specific rotation	+14°	−14°	0	?	?	0
melting point	?	?	56°	63°	63°	?

A3.13 **18** (*R*)-2, (*R*)-3,4-trihydroxybutanal
19 (*S*)-2, (*S*)-3,4-trihydroxybutanal
20 (*R*)-2, (*S*)-3,4-trihydroxybutanal
21 (*S*)-2, (*R*)-3,4-trihydroxybutanal

A3.14 **22** *threo*- **23** *erythro*- **24** *threo*-
(The structures **22**–**24** each represent a single enantiomer. The prefixes assigned apply equally correctly to the enantiomeric form or to the racemic mixture of the two.)

A3.15

CHO | CHO
HO—H | HO—H
HO—H | HO—H
CH_2OH | CH_2OH

HO, H, CHO, HO, H, CH_2OH

CHO, H, OH, HO, H, CH_2OH

A3.16

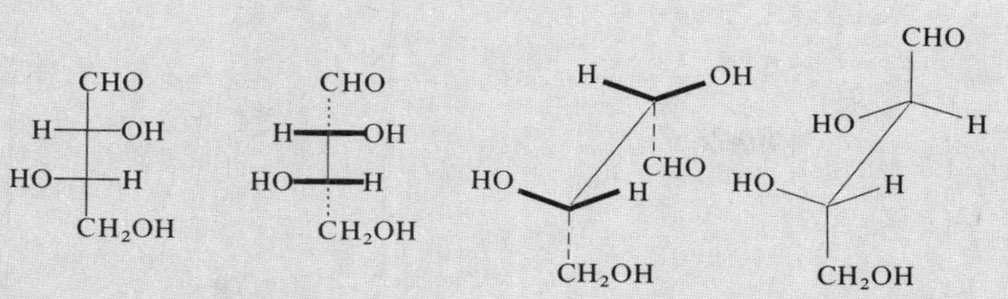

CHO
H—OH
HO—H
CH_2OH

CHO
H—OH
HO—H
CH_2OH

H OH
CHO
HO H
CH_2OH

CHO
HO H
HO H
CH_2OH

A3.17

CO_2H
H OAc
MeO H
Me

≡

AcO H
CO_2H
MeO H
Me

≡

CO_2H
AcO—H
MeO—H
Me

Ph
H Br
Br H
CO_2H

≡

Br H
Ph
Br H
CO_2H

≡

Ph
Br—H
Br—H
CO_2H

≡

CO_2H
H—Br
H—Br
Ph

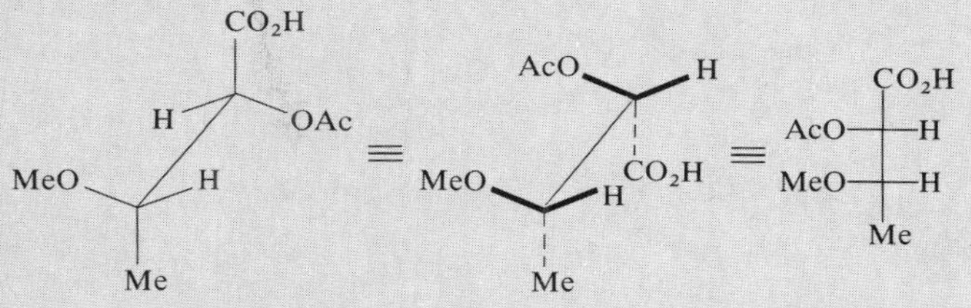

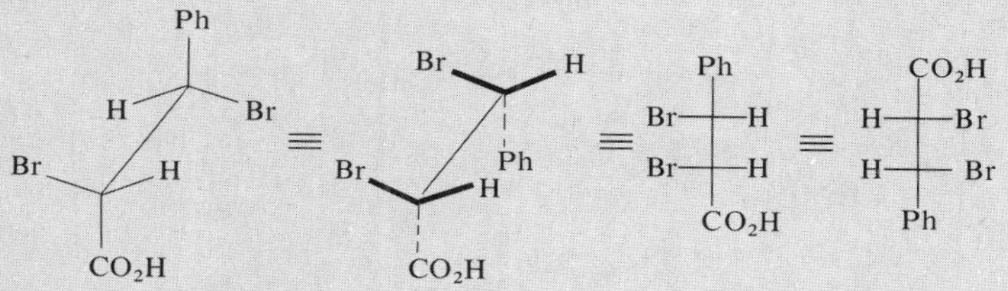

In the second example the CO_2H group C(1) should go to the top of the structure so, either the Fischer projection is rotated through 180° in the plane of the paper (the only permitted movement), or the original sawhorse projection could have been turned round to have the CO_2H system at the rear.

A3.18 Structures **36**, **38**, and **39** are the same: structure **37** is the same as **40** but these represent a different stereoisomer. (This result is most clearly derived by considering whether the ligands ABC and PQR, considered alphabetically, are present in a clockwise or anticlockwise order.)

A3.19

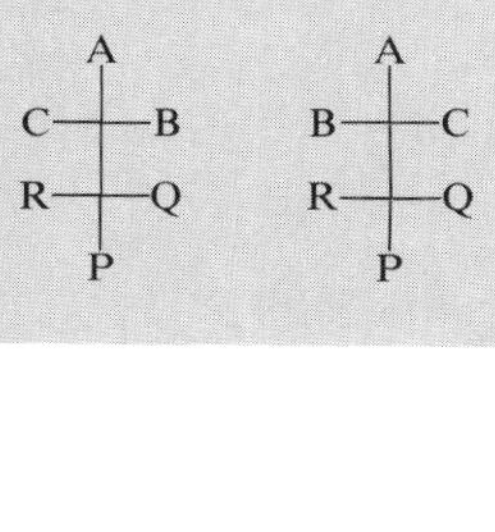

A
C—B
R—Q
P

A
B—C
R—Q
P

A3.20

CO_2H
H—Cl
Cl—H
Ph

CO_2H
Cl H
Cl H
Ph

Ph
H Cl
H Cl
CO_2H

or

CO_2H
Cl—H
H—Cl
Ph

CO_2H
H Cl
H Cl
Ph

Ph
Cl H
Cl H
CO_2H

A3.21

CO_2H
H—OH
HO—H
CO_2H
(**A**)

CO_2H
HO—H
H—OH
CO_2H
(**B**)

CO_2H
H—OH
H—OH
CO_2H
(**C**)

or

CO_2H
HO—H
HO—H
CO_2H
(**D**)

Tartaric acid exists in two optical forms (**A** and **B**), as a racemic mixture of these two, and as the *meso* (inactive) form represented by **C** or **D**.

meso-Compounds cannot be separated into enantiomeric forms and the racemic acid used by Pasteur was the racemic mixture of the two active (*threo*) isomers.

A3.22 enantiomers: **49** and **50**; **51** and **52**.
meso-compounds: **47** or **48** and **53** or **54**. [Each pair represents two equivalent Fischer projections which can be interconverted by rotation through 180° in the plane of the paper. Notice that each *meso*-compound contains a plane of symmetry.]

4

Conformation

4.1 Introduction

In those aspects of stereochemistry covered in the previous chapters it has been assumed, for the most part, that there is absence of rotation about double bonds in alkenes and related compounds (this is a requirement for *cis-trans*-isomerism) and that there is free rotation about single bonds. An exception to this final statement has already been reported, however, in the restricted rotation which exists in appropriately substituted biphenyls and in some compounds having a plane of chirality. Rotation about a bond joining two atoms does not necessarily have to be of these extreme types, i.e. no rotation or free rotation. Instead there is a continuum between the two. In compounds where rotation does not occur under normal temperature conditions isomeric molecules are said to differ in **configuration** as, for example, with maleic and fumaric acids. When the restriction to rotation is small the term **conformation** is applied. These two concepts merge and it is not possible to define the boundary between them. This is illustrated by the data in Table 1 which show the energy required to convert one conformer (or isomer) into another conformer (or isomer).

This chapter is mainly concerned with the nature and consequences of conformation in simple acyclic and cyclic molecules.

Table 1

	kJ mol^{-1}
ethane	12
cyclohexane	42
biphenic acid	63
2,2′ -di-iodobiphenyl	88
6,6′-dinitrobiphenic acid	125
but-2-ene	167

4.2 Acyclic compounds

A simple molecule like ethane may be represented by several sawhorse or Newman projections because of the possibility of free rotation about the single carbon-carbon bond. If this bond is rotated in 1° intervals then 360 projections can be drawn. Some of these will be identical because of the symmetry of this simple molecule and many will be very similar to one another. The manipulation of a molecular model will readily show two extreme situations represented by the sawhorse projections **1** and **2** or by the Newman projections **3** and **4**.

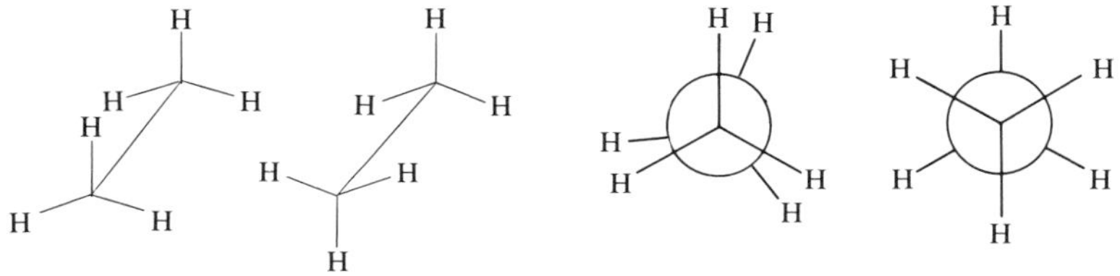

H H H H H H H H H H H H H H H H H H H H H H H H

(1) (2) (3) (4)

In projections **1** and **3** the alignments of the two sets of hydrogen atoms are the same and these atoms are said to be **eclipsed**. In projections **2** and **4** the hydrogen atoms are as far away from each other as they can be and they are said to be **staggered**. When the hydrogen atoms are eclipsed the angles between the atoms attached to the front and rear carbon atoms (the **torsional angle**) are zero. When the hydrogen atoms are staggered the torsional angle is 60°. Because of the symmetry of the ethane molecule there is only one staggered arrangement and only one eclipsed arrangement. The relative energies of these two systems will be considered after some less symmetrical systems have been examined.

Q4.1 Complete Newman projections such as **5** (with eclipsed substituents) and **6** (with staggered substituents) for butane ($CH_3CH_2CH_2CH_3$) in as many ways as possible in order to determine how many staggered and how many eclipsed conformations exist.

(5) (6)

Molecules less symmetrical than ethane can be written in additional conformations and, butane, for example, may exist in two eclipsed and two staggered arrangements (**7**–**10**). The two conformations with staggered arrangement of the substitutes have been designed as **gauche** (**9**) and **anti** (**10**) though these terms are increasingly replaced by others based on the torsional angle between two atoms or groups selected for reference purposes.

(7) (8) (9) (10)

In butane the two methyl groups are used for reference and the torsional angle between the two C-Me bonds can be varied between 0 and 360°. It is convenient to divide this range into six. The whole circle is divided into *syn* (together) and *anti* (against, opposite) semicircles (**11**) and into periplanar (about planar) and clinal (inclined) segments (**12**). Combination of these gives six segments.

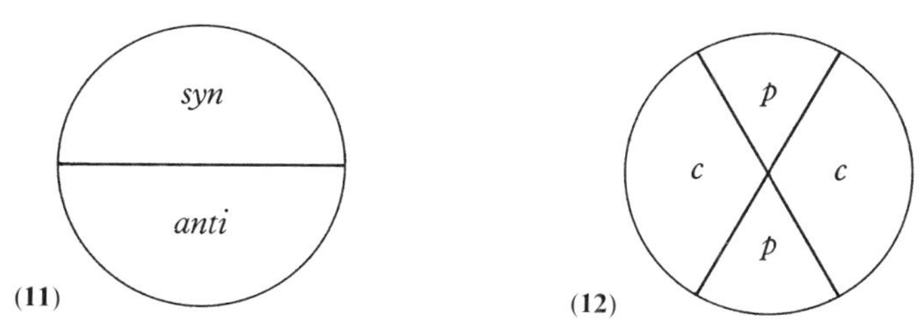

(11) (12)

Q4.2 Designate the six segments of **13** by appropriate symbols given that these are synperiplanar (*sp*), antiperiplanar (*ap*), synclinal (*sc*), and anticlinal (*ac*) and that any of these terms may be used more than once.

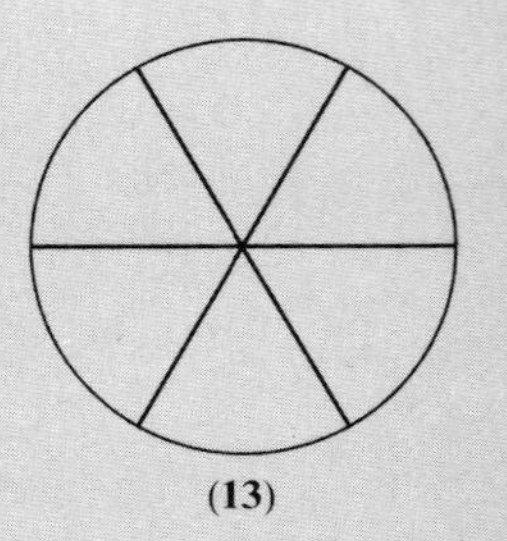

(13)

Q4.3 Using the two halogen atoms as reference groups describe the relation between these two in each of the Newman projections **14–17** by the terms (i) eclipsed or staggered and (ii) *sp*, *sc*, *ap*, *ac*.

(14) (15) (16) (17)

4.3 Energy differences between different conformations

The energy of the various conformations of the ethane molecule can be plotted against the torsional angle and it has been shown that this follows a sinusoidal curve (Fig. 4). The system has maximum energy when the hydrogen atoms are eclipsed (torsional angle = 0, 120, 240°) and the minimum energy when the hydrogen atoms are staggered (torsional angle = 60, 180, 300°). The difference between these two energy levels is about 12 kJ mol^{-1} and is too small to permit separation of isomeric molecules having eclipsed and staggered arrangements of the hydrogen atoms. The lower stability of the system with eclipsed atoms is believed to arise from interaction of the electron clouds which constitute the C-H bonds. This is referred to as **torsional energy** or **torsional strain**.

The position is a little more complex with butane since there are two staggered arrangements each with lower energy than the two eclipsed arrangements. The difference between the two conformers with staggered groups and the difference between the two conformers with eclipsed

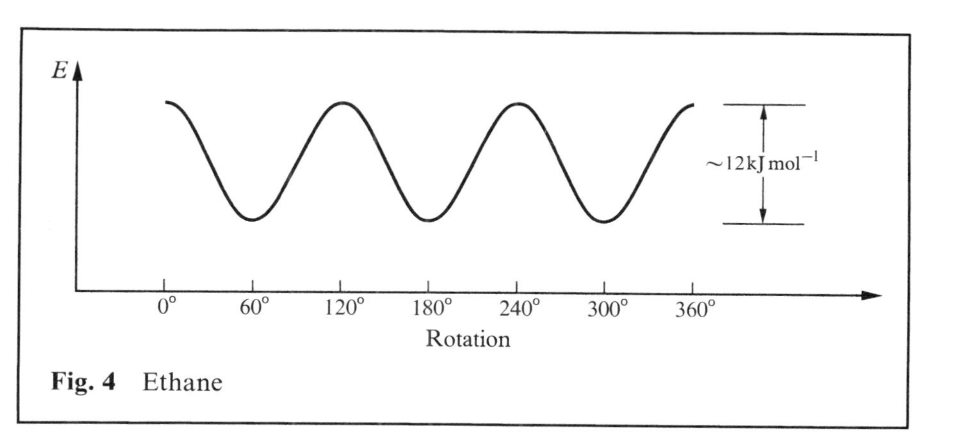

Fig. 4 Ethane

groups result from **steric interaction** between non-bonded groups (van der Waal's forces). These are considered to be insignificant between two hydrogen atoms, small between hydrogen and methyl, and larger between two methyl groups. The energy distribution shown in Fig. 5 is consistent with the view that torsional strain is usually greater than steric strain. These energy differences lead to the conclusion that, at 25°, the *ap* (69·0 per cent) and *sc* (30·4 per cent) forms predominate.

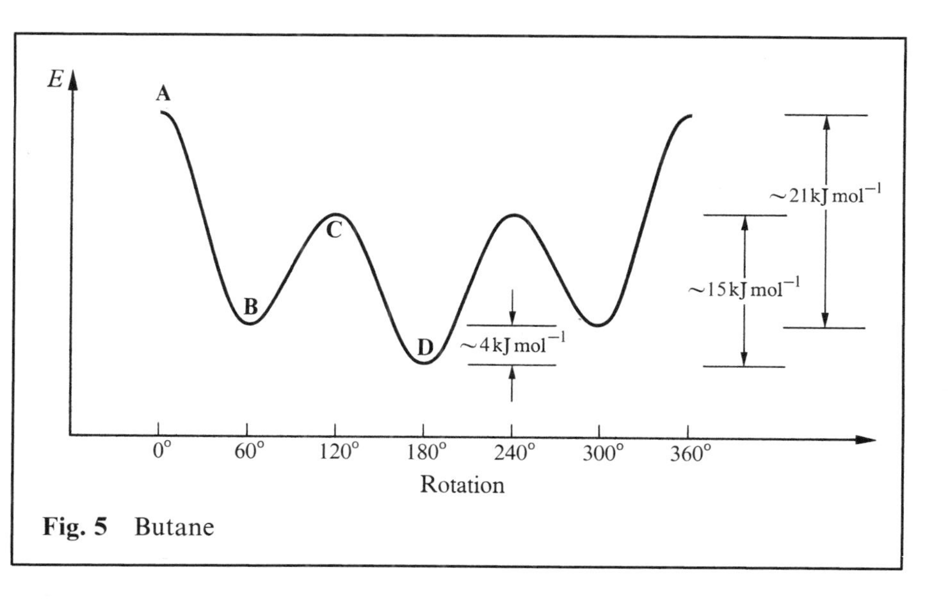

Fig. 5 Butane

Q4.4 Complete Newman projections of type **18**, rotating the groups attached to the rear carbon atom as required, to correspond to the points A, B, C, and D on Fig. 5, and designate them as *sp*, *sc*, *ap* or *ac* as appropriate.

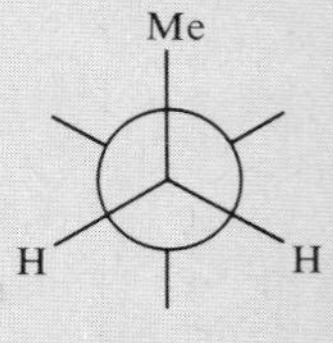

Me

H H

(**18**)

The conformation of an acyclic molecule depends on torsional strain, on steric strain, and also on hydrogen bonding and on the possibility of attractive or repulsive interaction between dipoles.

In 1,2-dibromoethane, for example, there are two carbon to bromine bonds both polarised in the manner $\overset{\delta+}{C}$—$\overset{\delta-}{Br}$ and since both bromine atoms carry a fractional negative charge they will tend to repel each other and to enhance the stability of the *ap* conformer (**19**). Hydrogen bonding, resulting in the formation of a quasi five-(or six-) membered ring, tends to hold the hydrogen-bonded atoms together. In ethane-1,2-diol other factors would stabilise the *ap* conformer (**20**) but hydrogen bonding stabilises the *sc* conformation (**21**).

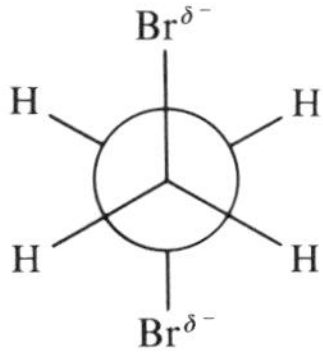

$Br^{\delta-}$ H H H H $Br^{\delta-}$

(**19**) (*ap*)

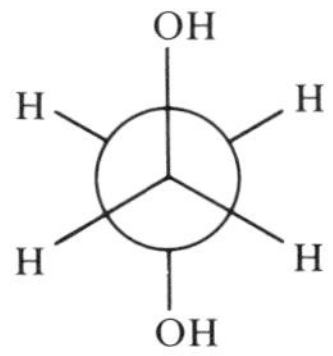

OH H H H H OH

(**20**) (*ap*)

intramolecular hydrogen bonding is not possible

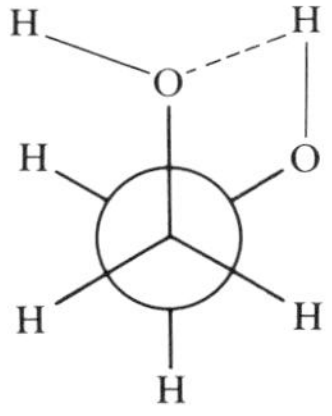

H O H O H H H H H

(**21**) (*sc*)

4.4 Cyclohexane

The most common cyclic compounds are those having six-membered rings. The conformation of these will be discussed first followed by some comments on polycyclic compounds and on rings of other sizes.

In saturated acyclic compounds each carbon-carbon bond has a valency angle of 109°28′ (the normal tetrahedral angle) or something fairly close to this. In ring systems this may not be possible. In the C_3 to C_6 cycloalkanes, for example, the bond angles will be as shown below if the

cyclopropane	cyclobutane	cyclopentane	cyclohexane
60°	90°	108°	120°

molecules are planar. Only cyclopentane approaches the normal tetrahedral angle (109°28′). The bond angle is smaller than this for the three- and four-membered rings and larger for rings with more than five bonded atoms. This last statement is only true for planar rings and rings with more than five carbon atoms can assume non-planar conformations in which the angle strain is wholly or largely eliminated. Four- and five-membered rings also assume non-planar conformations. This lack of planarity may increase angle strain but decreases other strains and produces, on balance, the most stable conformation.

Cyclohexane can exist in a chair form and in a number of flexible forms in all of which angle strain is largely eliminated. Among these the chair

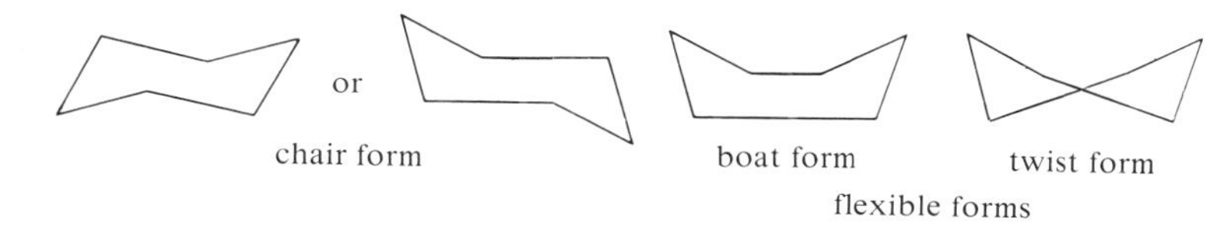

chair form boat form twist form
flexible forms

form is the most stable because torsional and steric strains are also minimised. (Nevertheless a flexible form may be the most stable conformation for some cyclohexane derivatives.)

Q4.5 Construct a molecular model of cyclohexane which may assume a chair, boat, or twist conformation.

(i) Examine each pair of adjacent carbon atoms in the chair form and notice that *all* groups are staggered.

(ii) Examine each pair of adjacent carbon atoms in the boat form and notice that some groups are staggered and some are eclipsed.

(iii) Examine each pair of adjacent carbon atoms in a twist form and notice that all groups are staggered.

(iv) Examine each structure for the possibility of interaction between non-bonded atoms other than those on adjacent carbon atoms. This is most pronounced between the flagpole groups

[C(1) and C(4)] in the boat form† and in the twist form and may occur to a lesser extent in the chair form between one of the groups attached to each of the C(1), C(3), and C(5) atoms and to each of the C(2), C(4), and C(6) atoms.

(v) Notice that in the chair form of cyclohexane three bonds project vertically above and three bonds project vertically below, the plane of the "chair-seat" (axial bonds) and the remaining six bonds project laterally from the plane of the ring (equatorial bonds).

(vi) Carefully label the six axial bonds and the six equatorial bonds in a model of the chair form of cyclohexane. Take the model and gently but firmly manipulate it so that the upward part of the chair becomes the downpart part and *vice versa*. Notice that as a result of this the axial bonds have become equatorial and the equatorial bonds have become axial.

If you have completed Question 4.5 you will have seen that (i) the chair form of cyclohexane is the most stable conformation because in this conformation angle strain, torsional strain, and steric strain have been minimised and (ii) in this form cyclohexane has six axial bonds and six equatorial bonds and that these are readily interconverted through ring-flip. The energy required for the change from one chair conformation to the other is not sufficient (~42 kJ mol^{-1}) to allow the two forms to be separated and ring-flip occurs ~10^6 times per second. At 25° only one molecule in 1000 is in the boat conformation, the remainder assume the chair form.

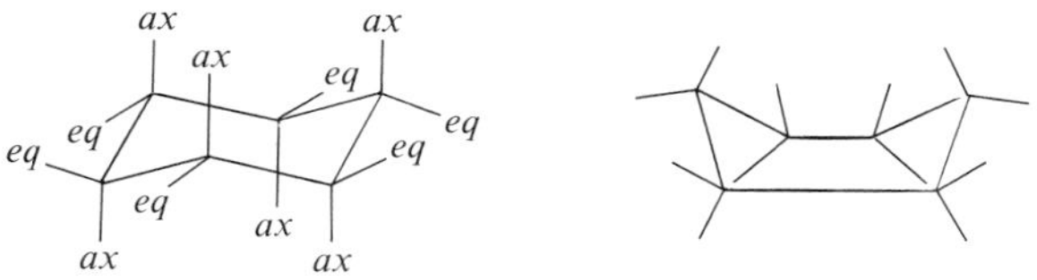

ax *ax* *ax* *eq* *eq* *eq* *eq* *eq* *eq* *eq* *ax* *ax* *ax*

At room temperature the various chair conformations of cyclohexane are equally populated and interchange so rapidly that all the hydrogen atoms have the same average environment and only one proton resonance signal is observed. In carbon tetrachloride solution, however, this peak

† The distance between the two hydrogen atoms in these positions is only 1·8 Å when the sum of their van der Waals radii is 2·4 Å.

gradually broadens as the temperature is lowered until, at temperatures below $-110°$, it separates into two peaks due to axial and equatorial hydrogen atoms respectively. Similarly the signal due to $C\underline{H}X$ protons of chloro- and bromo-cyclohexane is a single peak at room temperature, broadens at -40, and splits into two broad peaks at $-55°$. These belong to the two conformations in which the hydrogen is either axial or equatorial.

Axial protons generally absorb at slightly higher frequencies than the equatorial protons by a chemical shift of 0·1 to 0·7. Coupling constants for vicinal hydrogen atoms are usually J_{aa} 10–13 Hz, J_{ae} and J_{ee} 2–5 Hz.

In some ring compounds – such as pyranose sugar acetates – one of the two chair conformations is preferred at room temperature and separate signals are observed for non-equivalent axial and equatorial hydrogen atoms.

4.5 Substituted cyclohexanes

With substituents larger than hydrogen atoms it becomes significant that the equatorial positions are further apart than the axial positions and in a monosubstituted cyclohexane the equatorially substituted conformer is usually more stable than the axially substituted conformer. In other words, the molecule will spend more of its time in that conformation which places the substituent in an equatorial position. This disparity becomes more marked with increasing size of the substituent and the bulky t-butyl group is used to "lock" a cyclohexane ring in the chair conformation in which the t-butyl is equatorial.

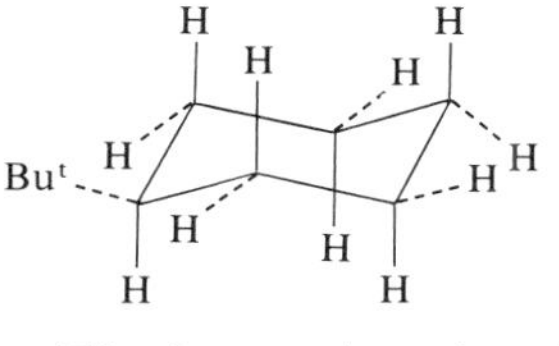

much more stable than

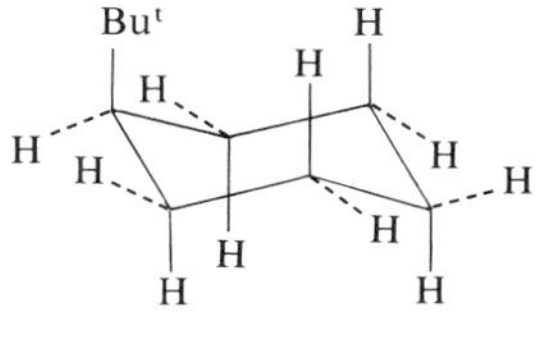

The interesting situation which arises with disubstituted cyclohexane derivatives is illustrated with a discussion of the dimethylcyclohexanes.

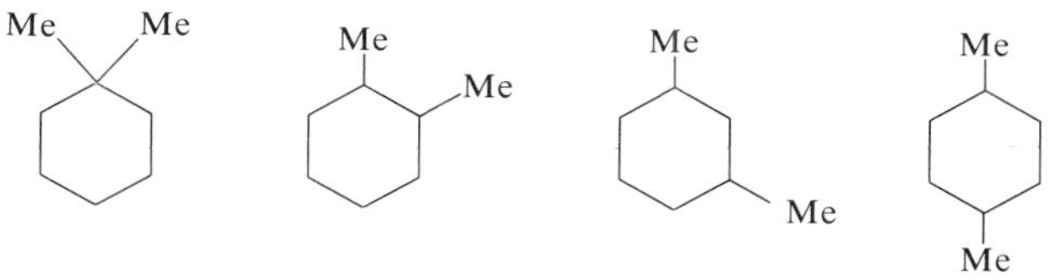

1,1- 1,2- 1,3- 1,4-
dimethylcyclohexanes

1,1-Dimethylcyclohexane exists only in one form and does not interest us here. The 1,2-, 1,3- and 1,4-dimethyl compounds each exist as *cis*- and *trans*-isomers and it is interesting to enquire (i) what are the possible conformations of each of the six isomers? (ii) for each isomer which is the more stable conformation? and (iii) for each dimethylcyclohexane is the *cis*- or the *trans*-isomer the more stable? There are two conformations for each of the six isomers and sometimes they are equivalent: where they are not the more stable conformer will contribute more to the structure under consideration.

Q4.6 (i) One conformation of *cis*-1,2-dimethylcyclohexane is shown in **22** and can be designated 1*e*2*a*. Draw and designate the conformer which results from **22** after ring-flip.

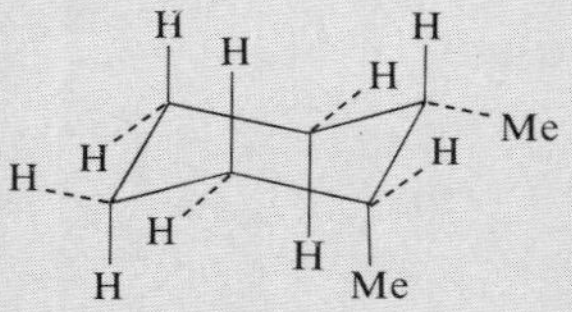

(22)

(ii) One conformation of *trans*-1,2-dimethylcyclohexane is shown in **23**. Designate the two substituents as axial or equatorial and draw and designate the conformer which results from **23** after ring-flip.

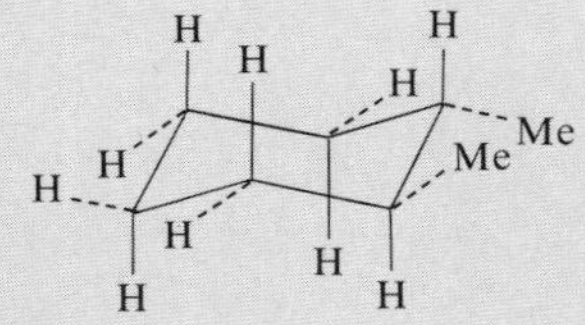

(23)

(iii) Given that the *cis*-isomer exists in the 1*e*2*a* or 1*a*2*e* conformations (which are equivalent and have the same energy) and the *trans*-isomer exists in the 1*e*2*e* or 1*a*2*a* conformation. Which of these isomers (*cis* or *trans*) is more stable?

Q4.7 Repeat the sequence of operations carried out in Q4.6 for the *cis*-(**24**) and the *trans*-(**25**) isomers of 1,3-dimethylcyclohexane and the *cis*-(**26**) and *trans*-(**27**) isomers of 1,4-dimethylcyclohexane. For each dimethylcyclohexane indicate which is the more stable isomer and what conformation makes the most significant contribution.

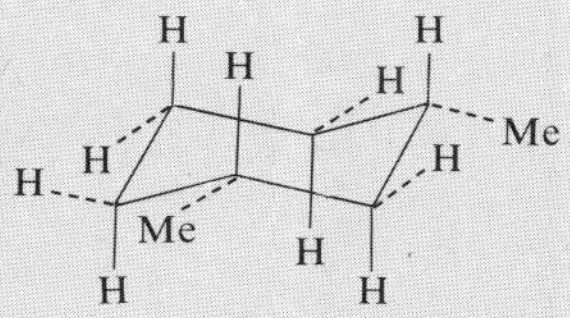

(**24**)

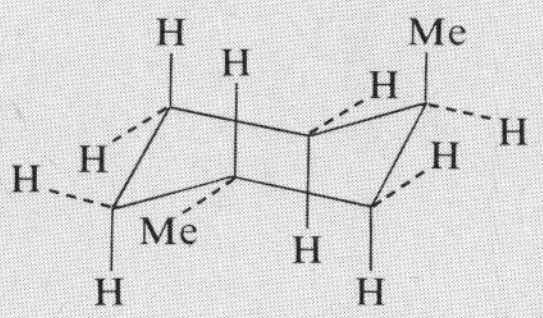

(**25**)

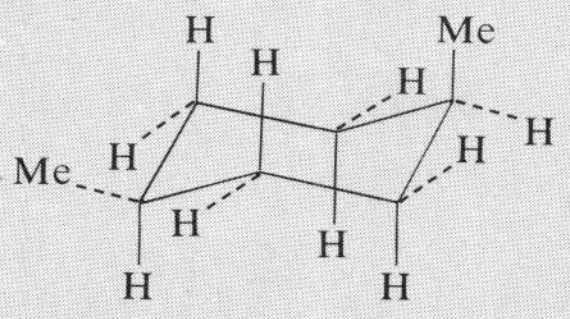

(**26**)

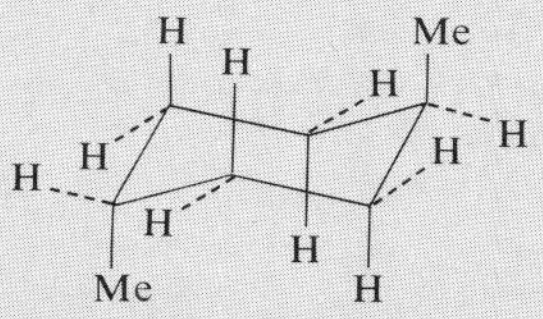

(**27**)

This type of study – known as **conformational analysis** – leads, even in this relatively simple case, to the unexpected conclusion that *cis*-1,3-dimethylcyclohexane is more stable than its *trans*-isomer. Experimental observations support this conclusion.

Other factors sometimes intervene and lead to a different conclusion. Two interesting examples are cited. In *trans*-1,2-dibromocyclohexane the diaxial conformation (**28**) makes a more significant contribution because in this form, and not in the diequatorial conformation, the carbon-bromine dipoles are opposed to each other. *cis*-Cyclohexane-1,3-diol is stabilised through being strongly hydrogen bonded and this is only possible when the hydroxyl groups are diaxial (**29**).

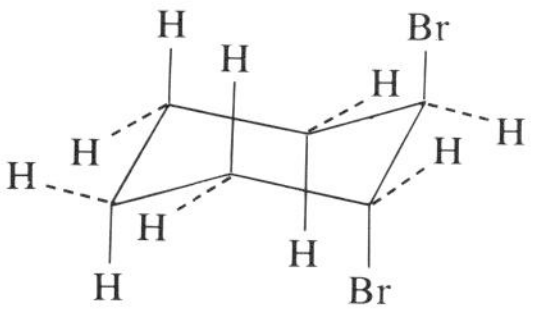

(**28**)

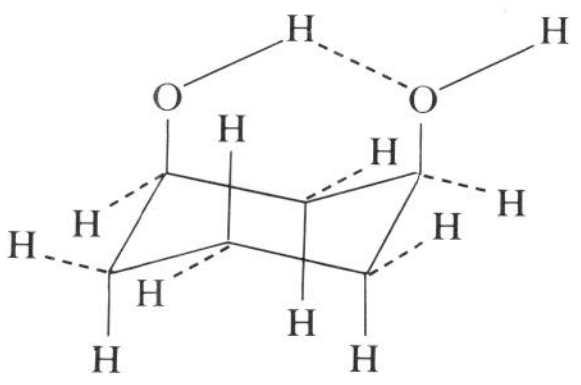

(**29**)

4.6 Cyclohexene

In cyclohexene four of the carbon atoms ($—CH_2CH{=}CHCH_2—$) are constrained into one plane. This molecule exists in half-chair (**30**) and half-boat (**31**) conformations of which the former is more stable. The hydrogen atoms in the half-chair form are almost in axial and equatorial positions and are designated quasi-axial (a') and quasi-equatorial (e').

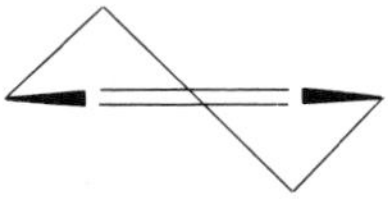

(**30**)

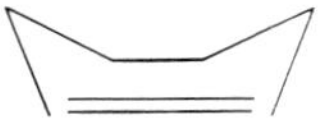

(**31**)

4.7 Decalin and other polycyclic systems

Decalin (decahydronaphthalene, **32**) has a molecular structure corresponding to two fused cyclohexane rings and may be considered as a derivative of 1,2-dimethylcyclohexane (**33**). Both these compounds exist in *cis* and *trans* forms and those for decalin are represented by structures **34** (*trans*) and **35** (*cis*). Six-membered rings are too small to permit di-axial ring fusion and therefore *trans*-decalin exists in only one double-chair conformation which is di-equatorially fused. *cis*-Decalin exists in two enantiomeric conformations which interconvert easily through a two-boat intermediate.

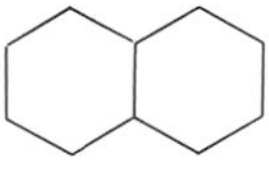

Me
Me

(**32**)

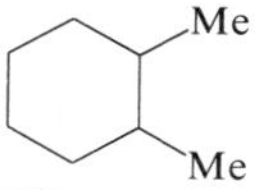

(**33**)

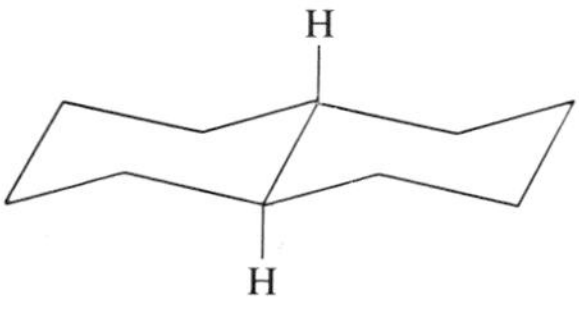

H
H

(**34**) *trans*

H
H
H
H

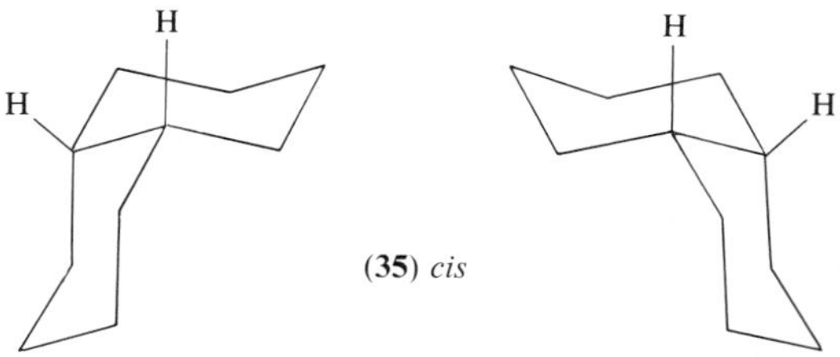

(**35**) *cis*

Q4.8 Set up a model of *trans*-decalin (**34**) and observe the following:

(i) both cyclohexane rings are in the chair form and all the groups are staggered,

(ii) the two hydrogen atoms at the bridging carbon atoms are axial with respect to both ring systems and are in a *trans* relation to each other,

(iii) the structure is rigid and cannot be changed to another all-chair conformation.

Q4.9 Set up a model of *cis*-decalin (**35** – either form) and observe the following:

(i) both cyclohexane rings are in the chair form and all the groups are staggered,

(ii) of the two hydrogen atoms at the bridging carbon atoms one is axial and the other is equatorial with respect to one ring and they have the reverse relationship with respect to the other ring; they have a *cis* relation to each other,

(iii) the structure is flexible and can undergo ring-flip to a second all-chair conformation, bonds that were axial become equatorial and *vice versa*.

It is a short step from decalin to important natural products, such as the steroids, which are based on perhydro-1,2-cyclopentanophenanthrene (**36**). The three six-membered rings exist in chair conformations but are

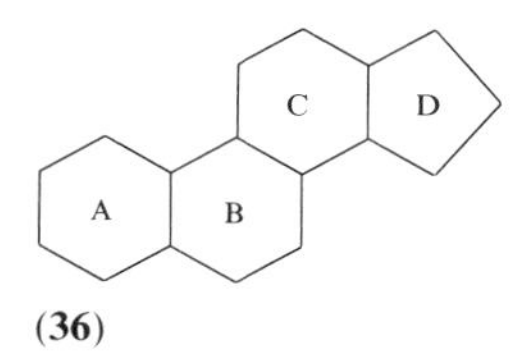

(**36**)

linked together in a rigid structure with axial and equatorial groups clearly defined. The appreciation of this has contributed greatly to our understanding of the chemical reactions of these compounds. There are two series of steroids based on the saturated compounds coprostane and cholestane. These differ only in the mode of linking between rings A and B which is *trans* in cholestane (a derivative of *trans*-decalin) and *cis* in coprostane (a derivative of *cis*-decalin). The remaining rings are *trans* linked.

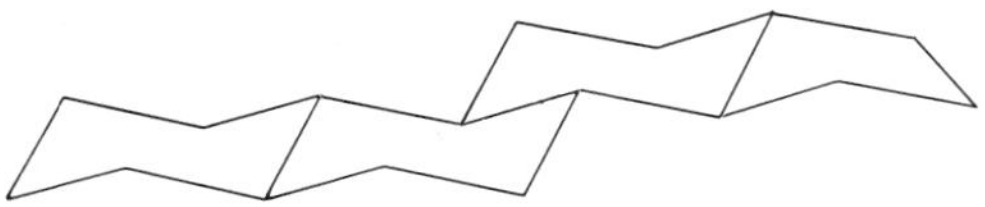

cholestane (*t*, *t*, *t*, *t*)

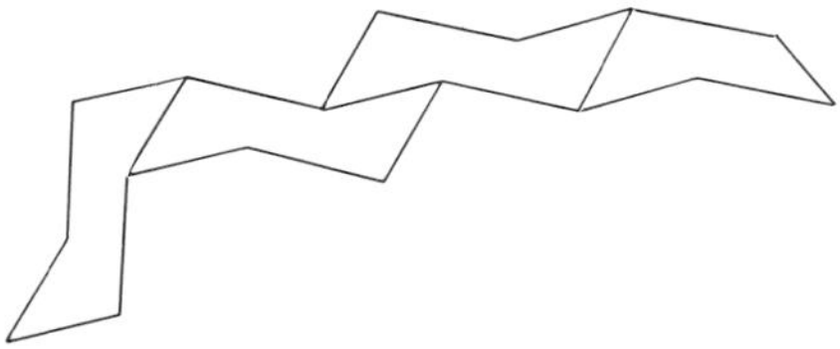

coprostane (*c*, *t*, *t*, *t*)

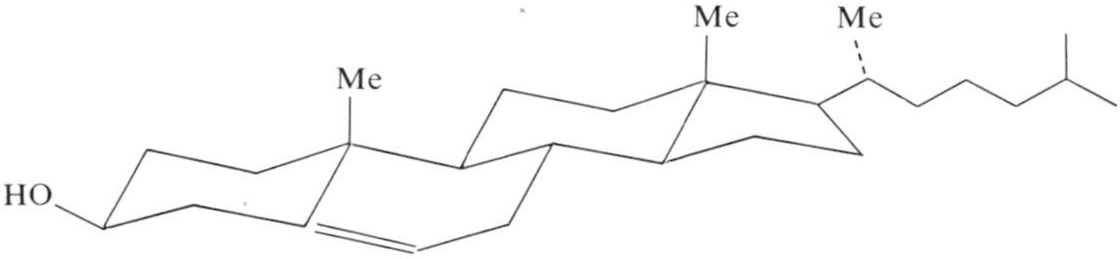

cholesterol

Cholesterol, the most common of all sterols, is a derivative of cholestane with the structure shown above.

The carbohydrates are another group of cyclic natural products where conformational analysis provides an additional insight into their structure and reactions.

α-D-glucopyranose

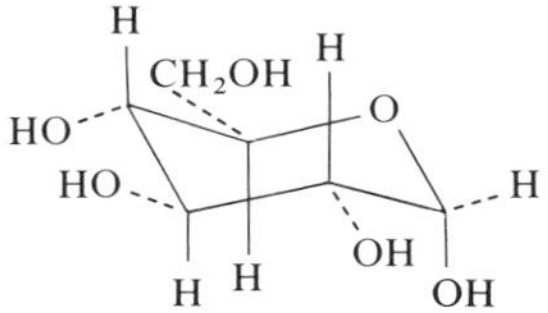

4.8 Influence of conformation on chemical properties

Groups in axial and equatorial positions differ in their thermodynamic stability and in their steric accessibility and these factors influence the chemical properties of molecules in differing conformations.

Equilibrium reactions will furnish the more stable isomer. For example, when ethyl 4-*t*-butylcyclohexanecarboxylate is heated with sodium ethoxide both the *cis*- and *trans*-isomers produce the same equilibrium mixture containing about 84 per cent of the more stable *trans*-isomer.

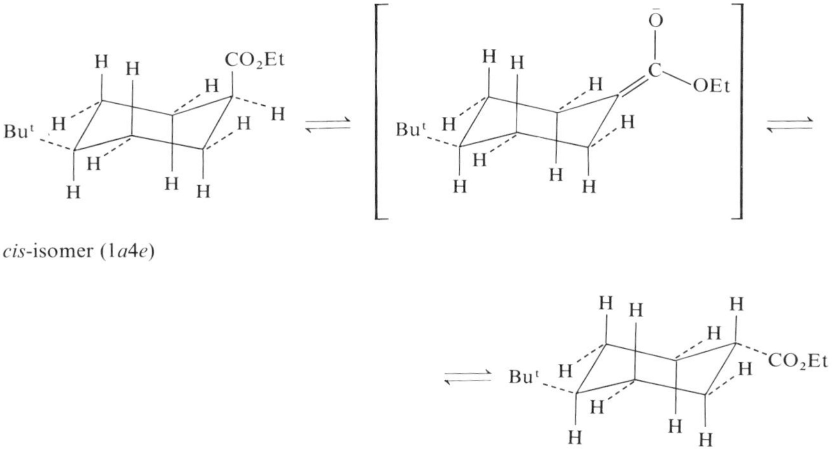

cis-isomer (1*a*4*e*)

trans-isomer (1*e*4*e*)

In kinetically controlled reactions consideration has to be given to the steric requirement of the transition state and to whether the reaction is controlled by its formation or its further reaction. In reactions in which the formation of the transition state is rate-controlling equatorial substituents are more reactive than similar axial substituents because of the lower steric restriction to the formation of the transition state in the equatorial position. The esterification of cyclohexanols and cyclohexanecarboxylic acids occurs more rapidly when the hydroxyl or carboxyl group is equatorial. The position is often reversed (as in the oxidation of cyclohexanols by chromic acid) when reaction is controlled by the decomposition of a transition state or intermediate because of the greater possibility of the relief of steric congestion, leading to steric acceleration, in the axial position. In bimolecular substitution reactions axial substituents are usually displaced more readily than equatorial substituents because the rearward approach of the attacking reagent (p. 90) is more hindered for equatorial than for axial substituents.

Elimination reactions are most commonly *trans* (Chapter 5) and require an antiperiplanar relationship between the two groups to be eliminated. This is only possible in the diaxial conformation and therefore *trans*-isomers, which can assume a diaxial conformation, react more readily than *cis*-isomers, which cannot.

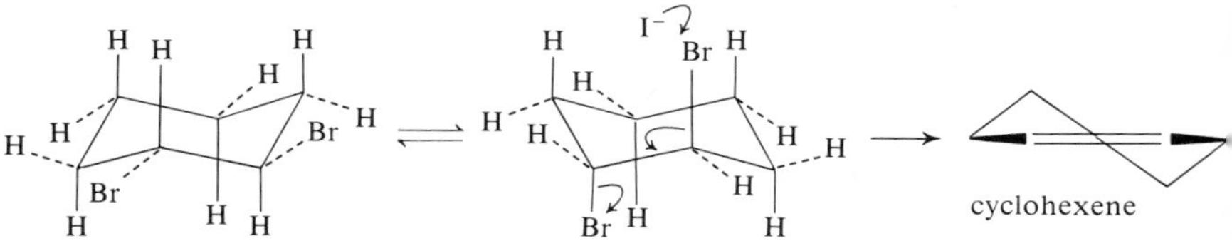

(1*e*2*e*)
trans-1,2-dibromocyclohexene

(1*a*2*a*)

cyclohexene

Pyrolytic eliminations require a (more or less) planar cyclic transition state which is best achieved with an equatorial-axial (*cis*) arrangement of groups.

4.9 Other ring systems

We have said a great deal about six-membered rings because compounds containing this system are easily prepared and provide rewarding study. What about rings of other sizes?

The stability of a ring system depends on several factors and the most stable form of the ring will be that in which these are balanced to give the minimum energy. The factors to be considered are:

(i) Angle strain resulting from the deviation from the normal tetrahedral angle (109°28′).

(ii) Torsional strain resulting from non-bonded interactions between groups attached to adjacent atoms.

(iii) Steric strains resulting from non-bonded interactions between groups on non-adjacent atoms as, for example, between 1,3-diaxial substituents in cyclohexane derivatives. This becomes more significant in medium sized rings where transfer of groups across a ring is frequently observed (transannular reactions).

The ease of formation of a ring system will also depend on an additional factor. Most synthetic procedures involve an intramolecular reaction between two reactive centres in a single molecule and the ease of this reaction depends, in part, on how much of their time the two reactive centres spend in close propinquity. This will decrease with increasing separation of the two reactive centres since the number of possible conformations which an acyclic system may assume will increase with its length. It is for this reason that three-membered rings, though subject to considerable strain, are nevertheless formed more easily than four-membered rings.

Four- and five-membered rings are not planar since puckering results in some reduction of torsional strain. Rings larger than cyclohexane are, like this molecule, flexed to reduce both angle strain and torsional strain though this may result in significant transannular interactions. Possible conformations for cyclo-octane and cyclododecane are shown.

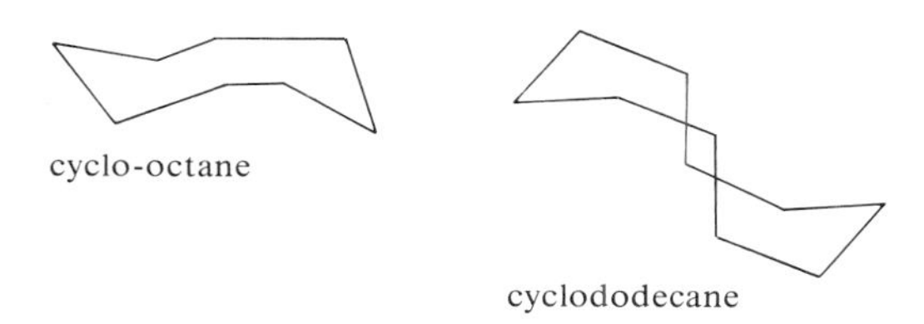

4.10 Stereochemistry of compounds of elements other than carbon

So far in this book attention has been focussed almost entirely on organic compounds. This situation arises from the facts that stereoisomerism is widely observed among organic compounds, that its nature has been recognised for over a century, and that methods of producing chiral molecules are most highly developed among carbon compounds. Nevertheless, stereochemistry is not confined to carbon compounds and the discussion will now be extended to other elements.

We begin with some general comments on molecular geometry, i.e. the spatial disposition of ligands around a central atom. The atom A may be attached to several ligands to give compounds AL_n and such compounds may also contain lone pairs E associated with A which, along with the bonding electron pairs, influence molecular geometry. These compounds should be considered as AL_nE_m. Molecular geometry depends on the number of valence shell electron pairs (bonding *and* lone) possessed by A.

The molecular shapes which a compound AL_n may assume are set out in Table 2 for values of n between 2 and 8. The previous account of the stereochemistry of organic compounds is really an extended discussion of one section of Table 2, *viz.* compounds in which A has a coordination number of four and in which the ligands assume a tetrahedral distribution around A. We now examine some other systems and consider the possibilities of stereoisomerism and the stability of stereoisomers. It will be recalled that a compound is potentially enantiomeric if its molecular structure

Table 2

Number of ligands	*Shape of molecule*
2	linear angular
3	planar pyramidal
4	tetrahedral square planar
5	trigonal bipyramidal square pyramidal
6	octahedral
7	pentagonal pyramidal
8	cubic square antiprismatic dodecahedral, etc.

cannot be superimposed on its mirror image and that stable enantiomers will exist only if there is an appropriate energy barrier between the two forms.

Carbon is not the only atom with a coordination number of four and a tetrahedral arrangement of ligands. Other elements share these properties and the possibilities of chirality, discussed at length in Chapter 3, apply equally to appropriate derivatives of other elements. It is not surprising, therefore, that optically active compounds of silicon are known. Derivatives of Group 5 elements which can be written as AL_3E or $\overset{+}{A}L_4$ are, however, of greater interest. In both of these classes of compounds the central atom is linked to four groups – one of which may be a lone pair – and if these are tetrahedrally arranged then there should be some similarity between these and analogous carbon compounds in respect of stereochemical possibilities. It is thus not surprising that ammonium and phosphonium salts exist in non-superimposable (chiral) forms and the oxide **37** and salts of the ions **38** and **39** in which nitrogen or phosphorus is attached to four different ligands have been separated into enantiomeric forms.

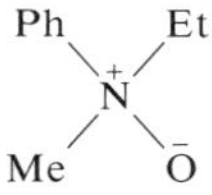

Ph Et $\overset{+}{N}$ Me $\bar{O}$

(**37**)

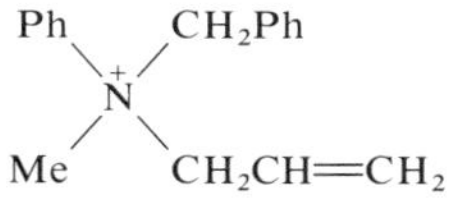

Ph CH_2Ph $\overset{+}{N}$ Me $CH_2CH{=}CH_2$

(**38**)

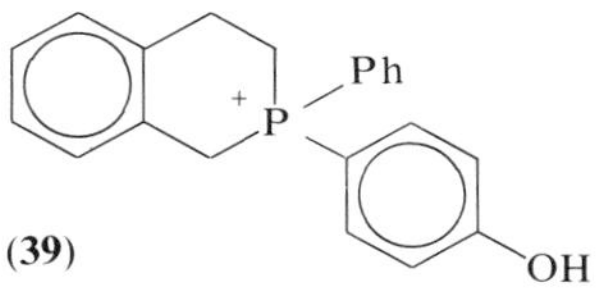

$\overset{+}{P}$ Ph OH

(**39**)

Trivalent derivatives of nitrogen, phosphorus, arsenic, and sulphur exist in pyramidal form though this can be considered as tetrahedral if the lone pair is counted as a fourth ligand. Among nitrogen derivatives inversion of pyramidal forms occurs very easily via a planar transition state so that the separation of enantiomers (**40** and **41**) is not possible except in special cases.

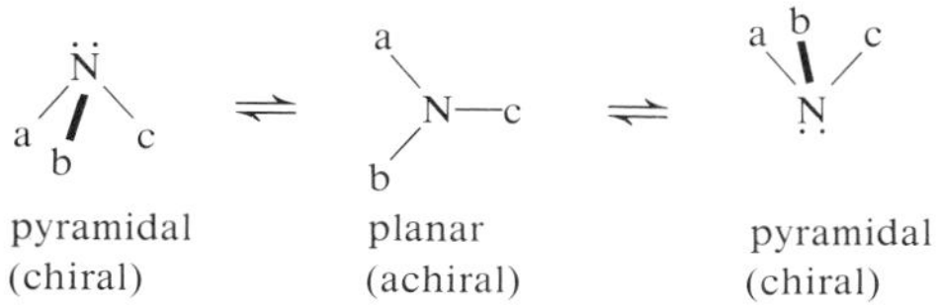

pyramidal (chiral) ⇌ planar (achiral) ⇌ pyramidal (chiral)

(**40**) (**41**)

The ammonia molecule is known to be pyramidal with a height of 0·30 Å but inversion occurs at a rate of $2{\cdot}4 \times 10^{10}$ cycles per second, the energy of activation for this change being only 25 J mol^{-1}. This reversible inversion can be prevented by holding the nitrogen pyramid in a rigid system and compound (**42**) in which both nitrogen atoms are part of two ring systems has been isolated in an optically active form.

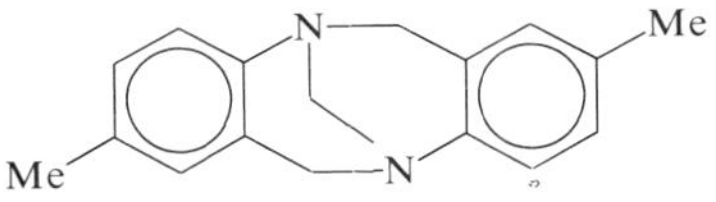

(**42**)

Phosphorus, arsenic, and sulphur derivatives have a higher energy of activation and the rate is slow enough to permit separation of the dia-stereoisomeric camphorsulphonates of the sulphonium ions **43** and **44**.

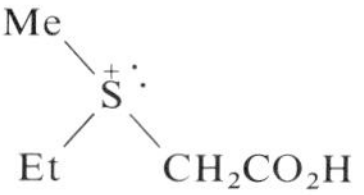

Me, Et, S^+, CH_2CO_2H — (**43**)

Me, Et, S^+, CH_2COPh — (**44**)

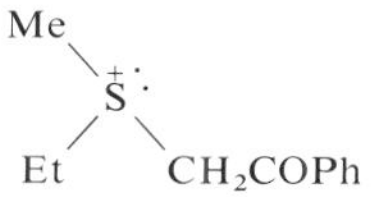

Compounds in which the central atom has a coordination number of five, including, for example, many derivatives of phosphorus, present new stereochemical features. Such compounds exist mainly as trigonal bi-pyramids and less commonly as square pyramids. Spectroscopic studies show that compounds like PF_5 or $Fe(CO)_5$ exist as trigonal bipyramids: three of the ligands (L_e, equatorial) are in the same plane as the central atom and attached to it by bonds at angles of 120° and the remaining two ligands (L_a, axial, apical) are positioned above and below the equatorial plane. It is now accepted that molecules of the type AL_4E, AL_3E_2, and AL_2E_3 will have shapes based on a trigonal bipyramid, that lone pairs

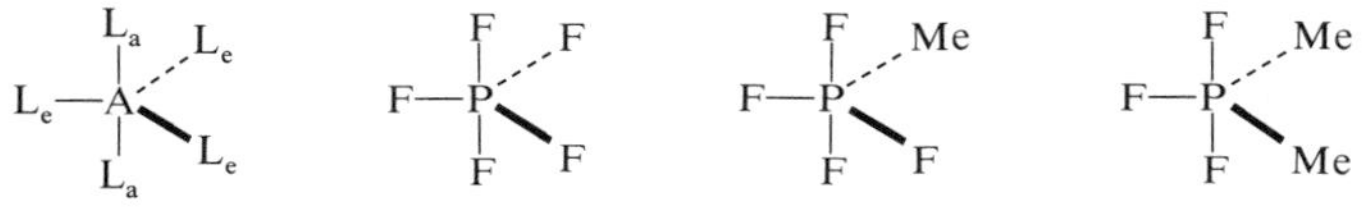

(**45**) (**46**)

occupy equatorial positions, as do atoms which are multiply bonded to A, and that when there is more than one type of ligand the least electronegative occupy equatorial positions. Thus phosphorus halides of the types PF_4R and PF_3R_2 (R=alkyl or aryl groups) do not show the stereoisomerism which is theoretically possible but exist solely in those forms in which the organic groups occupy equatorial positions (**45** and **46**).

All this points to ligands of two types: three of which are equatorial and two of which are axial. Nevertheless the careful nmr study of PF_5, even at low temperature, does not give two signals corresponding to axial and equatorial fluorine atoms. The compound Me_2PF_3, however, does show two ^{19}F signals at low temperature. The unexpected result for PF_5 is usually discussed in terms of stereochemically non-rigid structures; a term applied to molecules that undergo rapid intramolecular rearrangement. This facile internal rearrangement of the trigonal bipyramid has been explained by a number of mechanisms and one that finds common acceptance is pseudorotation first proposed by Berry. If the L_1AL_2 angle, normally 180°, is reduced slightly and the L_4AL_5 angle, normally 120°, is

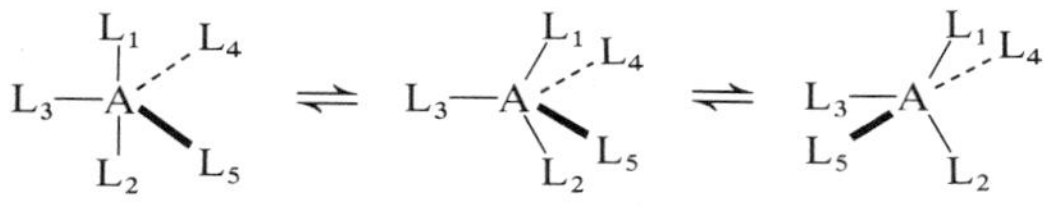

trigonal bipyramid ⇌ square pyramid ⇌ trigonal bipyramid

increased slightly then a square pyramid results and this can either revert to the original trigonal bipyramid or change to a second trigonal bipyramid. The consequence of this inversion is that two equatorial ligands have become axial and two axial ligands have become equatorial. The equatorial bond not involved in this change is described as the "pivot bond" but since any two of the three equatorial ligands are involved in pseudorotation, all five ligands become equivalent. The problem is more complex when pairs of ligands are part of the same cyclic system.

This non-rigidity means that even in apparently chiral structures, enantiomeric compounds are unlikely to be isolated.

Compounds of atoms with a coordination number of six show octahedral geometry. The octahedron, like the tetrahedron, is a stable (rigid) configuration and many compounds are chiral and are known in their enantiomeric forms. Examples are to be found among the enantiomeric metal complexes such as **47** and **48**.

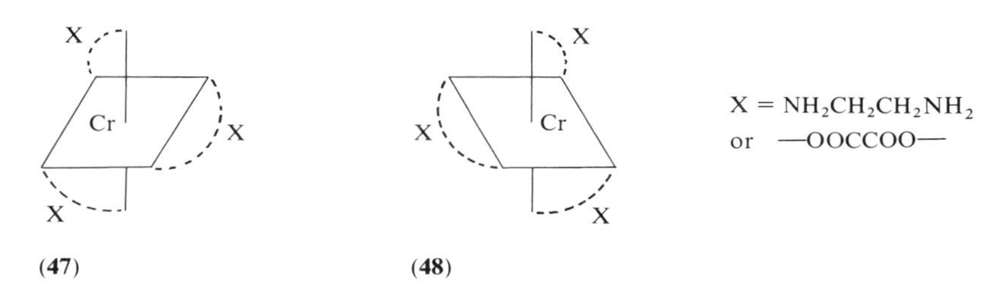

(**47**) (**48**)

Answers

A4.1 There are two eclipsed arrangements (**7** and **8**) and two staggered arrangements (**9** and **10**). Any additional projections must be identical with one of these.

A4.2

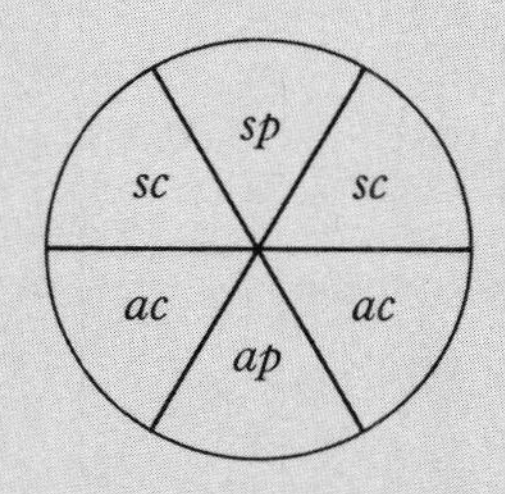

A4.3 **14** eclipsed (*sp*) **15** staggered (*sc*)
16 staggered (*ap*) **17** eclipsed (*ac*)

A4.4

(**18A**) (*sp*) (**18B**) (*sc*) (**18C**) (*ac*) (**18D**) (*ap*)

A4.6 (i) (ii) 1*e*2*e*

1*a*2*e* 1*a*2*a*

(iii) The *trans*-isomer (1*e*2*e*$\gg$1*a*2*a*) is more stable than the *cis*-isomer (1*e*2*a*$=$1*a*2*e*).

A4.7

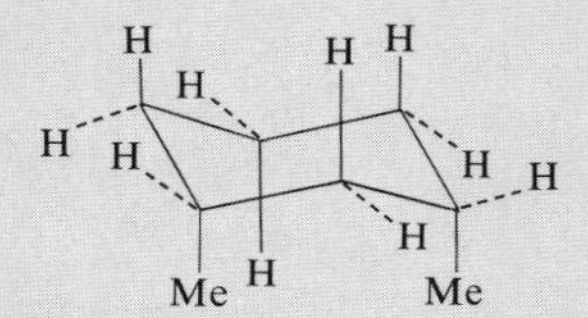

1,3-dimethylcyclohexane: *cis*, 1*e*3*e* and 1*a*3*a*; *trans*, 1*a*3*e* and 1*e*3*a*: the *cis*-isomer (1*e*3*e*$\gg$1*a*3*a*) is more stable than the *trans*-isomer (1*a*3*e*$=$1*e*3*a*).

1,4-dimethylcyclohexane: *cis*, 1*a*4*e* and 1*e*4*a*; *trans*, 1*a*4*a* and 1*e*4*e*: the *trans*-isomer (1*e*4*e*$\gg$1*a*4*a*) is more stable than the *cis*-isomer (1*a*4*e*$=$1*e*4*a*).

5

Dynamic stereochemistry

Earlier chapters in this book have been concerned with static stereochemistry; that is, with the possibilities of stereoisomerism in individual compounds in the ground state. The stereochemical nature of a chemical reaction is of the utmost importance when reactant and/or product exhibit stereoisomerism, and this final chapter is devoted to the stereochemical consequences of some of the more important types of chemical reactions: in particular addition, elimination, substitution, and rearrangement reactions. Information of this kind is required when planning all except the simplest of synthetic sequences and there is a continuing demand among chemists for reactions of increasing stereospecificity.

Enzymic processes usually occur with a high level of stereospecificity; for example, in the reductive chain-elongation process by which fatty acids are made in nature the following steps occur:

$$\mathrm{RCOH_2COEnz} \longrightarrow \underset{(\mathbf{1})}{\mathrm{RCH(OH)CH_2COEnz}} \longrightarrow \underset{(\mathbf{2})}{\mathrm{RCH{=}CHCOEnz}} \longrightarrow \mathrm{RCH_2CH_2COEnz}$$

Each reaction requires an enzyme and appropriate cofactors and is highly stereospecific. Thus the hydroxyacyl compound (**1**) is entirely the (*R*)-3 isomer and this is dehydrated only to the *trans*-2-acyl derivative (**2**). The 2*c*, 3*c* or 3*t* isomers, if introduced into the system, do not undergo the final reduction.

5.1 Addition reactions

Many important reactions of alkenes and alkynes are addition processes which can be written in a general form such as:

$$\mathrm{RC{\equiv}CR'} \xrightarrow{X_2} \mathrm{RCX{=}CXR'}$$

(**3**) (**4**)

$$\mathrm{RCH{=}CHR'} \xrightarrow{X_2} \mathrm{RCHXCHXR'}$$

(**5**) (**6**)

The alkyne (**3**) exists in only one form but the alkenes (**4** and **5**) exist in *Z* or *E* (*cis* and *trans*) forms and the addition product (**6**), with two non-identical chiral centres, exists in four enantiomeric forms which can be paired into two racemates (*threo* and *erythro*).

Addition reactions which are not stereospecific are of little synthetic value except with simple non-stereoisomeric molecules.

Q5.1 Consider the hydroxylation reaction formulated below and answer the attached questions:

$$\mathrm{RCH{=}CHR'} \xrightarrow{\text{hydroxylation}} \mathrm{RCH(OH)CH(OH)R'}$$

(i) Formulate the two stereoisomeric alkenes and designate each as *E* or *Z*.

(ii) Write Fischer projections for the four enantiomers of the diol (place the R group uppermost) and designate them as *threo* or *erythro*.

The answers to Q5.1 raise a number of further questions. Will addition to alkynes produce the *Z* or *E* alkene or a mixture of these? Do the addition reactions of alkenes furnish *threo* or *erythro* adducts or both and to what extent does the answer to this depend on the configuration of the substrate and the nature of the addition reaction?

Although addition reactions may occur by different mechanisms only two stereochemical consequences have to be considered. The two groups may be added to the *same* side of the unsaturated system or to *opposite* sides. These are known as *cis* addition (suprafacial) and *trans* addition (antarafacial) respectively. The substrate, the addition process, and the products are then related as shown in Scheme 1.

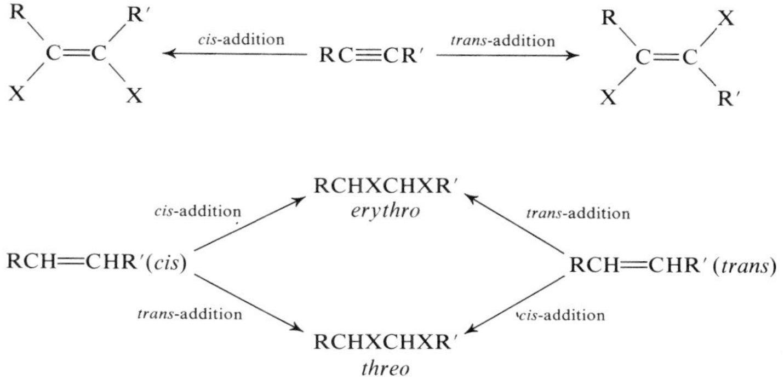

R(X)C=C(R')(X) ← *cis*-addition — RC≡CR' — *trans*-addition → R(X)C=C(X)(R')

RCH=CHR'(*cis*) — *cis*-addition → RCHXCHXR' *erythro* ← *trans*-addition — RCH=CHR' (*trans*)

RCH=CHR'(*cis*) — *trans*-addition → RCHXCHXR' *threo* ← *cis*-addition — RCH=CHR' (*trans*)

Scheme 1 Stereochemical relationships between alkynes and alkenes and their addition products.

The statements formulated in these sequences will be established shortly but first we examine their significance. An *erythro* adduct results either from a *cis*-alkene by *cis*-addition or from a *trans*-alkene by *trans*-addition. Conversely the *threo* adduct results from the *cis*-alkene by *trans*-addition or from the *trans*-alkene by *cis*-addition. These observations are significant when there are two ways of adding X_2 to an alkene, one involving *cis*-addition and the other involving *trans*-addition. If there is only one procedure available then a particular adduct can only be obtained from the appropriate alkene. For example, if only a *cis*-addition procedure is known then the *erythro* adduct must be obtained from the *cis*-alkene and the *threo* adduct must be obtained from the *trans*-alkene. Obviously this type of information must be fully appreciated before embarking on a synthetic sequence.

Q5.2 The following sequence shows how the addition of X_2 to a *cis*-alkene in a *cis* manner from above the molecule gives an *erythro* adduct.

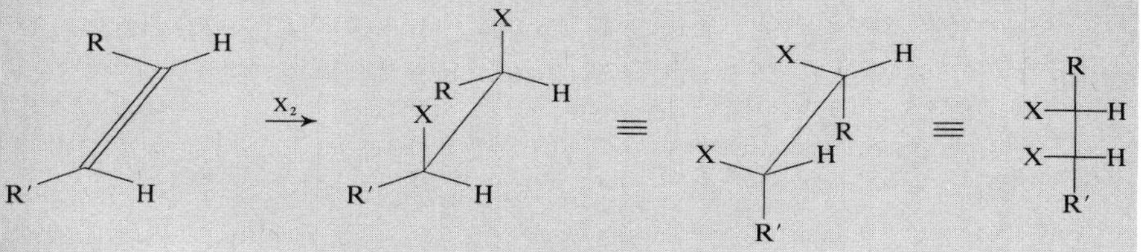

X_2 → ≡ ≡

Remember that in a Fischer projection the groups at the top and bottom of the projection lie *below* the plane of the paper and those to the right and left lie *above* the plane of the paper.

In general, *cis*-addition may occur equally well from below the molecule. Show by a similar sequence that this gives the other *erythro* enantiomer so that the addition product is the racemic *erythro* form.

Similar exercises, on paper or with models, will demonstrate the correctness of other relationships summarised in Scheme 1.

5.2 Hydrogenation

Catalytic hydrogenation is usually a *cis*-addition and hydrogenation of an alkyne proceeds via the *cis*-alkene to the alkane which is not stereoisomeric. The reaction can be halted at the *cis*-alkene stage if a partially

$$RC{\equiv}CR' \xrightarrow{H_2,\ Pd} \underset{cis}{RCH{=}CHR'} \xrightarrow{H_2,\ Pd} RCH_2CH_2R'$$

poisoned palladium catalyst is employed and this is the basis of one of the best methods of preparing alkenes having the *cis*-configuration.† It is usually better to use stereospecific reactions which give the desired isomer than to have to separate *cis*- and *trans*-isomers produced by some non-stereospecific process.

Alkynes and alkenes are also reduced by hydrazine (the reactive species is probably di-imide, N_2H_2). This reaction is not easily controlled to give

$$RC{\equiv}CR' \xrightarrow{N_2H_4} RCH{=}CHR' \xrightarrow{N_2H_4} RCH_2CH_2R$$

$$RCH{=}CHR' \xrightarrow{N_2D_4} RCHDCHDR'$$

the alkene but it has been used to prepare alkanes with deuterium atoms on adjacent carbon atoms. If prepared from the *cis*-alkene the dideuterio-alkane will be the *erythro* isomer as this also is a *cis*-addition, reaction

† The reduction of an alkyne with sodium and liquid ammonia is a *trans*-addition and gives the *trans*-alkene. This is a useful preparative route for the stereospecific formation of *trans*-alkenes. (For an example see p. 12.)

occurring through a quasi-cyclic intermediate. Conversely the *threo* adduct can be prepared from the *trans*-alkene.

$$\text{C(D)=C(D)} \cdots \text{N=N} \longrightarrow -\text{C(D)}-\text{C(D)}- + N_2$$

5.3 Bromination and hydroxylation

Other important addition reactions include halogenation (especially bromination), which is usually a *trans*-addition, and hydroxylation, which

$$\text{RCHBrCHBrR}' \xleftarrow{\text{bromination}} \text{RCH=CHR}' \xrightarrow{\text{hydroxylation}} \text{RCH(OH)CH(OH)R}'$$

may be *cis* or *trans* depending on the choice of reagent. In summary, halogenation and hydroxylation with peracids are *trans*-addition processes and hydroxylation with potassium permanganate or with osmium tetroxide are *cis*-addition processes.

All of these reactions start with an initial *cis*-addition leading to a three- or five-membered ring followed by ring opening which may occur with or without inversion. *cis*-Addition followed by inversion is equivalent overall to *trans*-addition.

overall stereochemistry of addition process

$$\text{RCH=CHR}' \xrightarrow[\textit{cis}\text{ addition}]{Br_2} \left[\overset{Br^+}{\text{RCHCHR}'}\right] \xrightarrow[\text{inversion}]{Br^-} \text{RCHBrCHBrR}' \qquad \textit{trans}$$

$$\text{RCH=CHR}' \xrightarrow[\textit{cis}\text{ addition}]{\text{peracid}} \overset{O}{\text{RCHCHR}'} \xrightarrow[\text{inversion}]{H^+,\ H_2O} \text{RCH(OH)CH(OH)R}' \qquad \textit{trans}$$

$$\text{RCH=CHR}' \xrightarrow[\textit{cis}\text{ addition}]{KMnO_4} \text{RCH(O)}-\text{CHR}'\text{(O)}>\text{Mn}(=\text{O})\bar{\text{O}} \xrightarrow{\text{hydrolysis}} \text{RCH(OH)CH(OH)R}' \qquad \textit{cis}$$

$$\text{RCH=CHR}' \xrightarrow[\textit{cis}\text{ addition}]{OsO_4} \text{RCH(O)}-\text{CHR}'\text{(O)}>\text{Os}(=\text{O})_2 \xrightarrow{\text{hydrolysis}} \text{RCH(OH)CH(OH)R}' \qquad \textit{cis}$$

Q5.3 Complete the following sequence to demonstrate that bromination is a *trans*-addition and that a *trans*-alkene yields the *erythro* dibromide.

R H

trans-alkene

H R′

Br_2

(*cis*-addition)

Br^+

+

Br

Br^-

(inversion)

Br Br

+ Br Br + + Br Br

Br Br

R R

+ *erythro* dibromide

R′ R′

5.4 Elimination reactions

Elimination reactions such as dehydration, dehydrohalogenation, dehalogenation, and the elimination reactions shown by quaternary ammonium compounds and certain esters are important for the preparation of alkenes and alkynes. Since both substrate and product may be capable of existing in stereoisomeric forms the stereochemical nature of these reactions becomes significant. In many respects the stereochemistry of elimination processes is the reverse of the stereochemistry of addition processes.

Both radical and polar elimination reactions are known and the latter occur by unimolecular (E1) or bimolecular (E2) mechanisms. The steric nature of the radical and E1 polar processes is difficult to control and to predict but E2 processes are more stereospecific and both *cis* and *trans* elimination processes are known. The latter are more common and will be discussed first.

Trans elimination of two atoms or groups from adjacent (vicinal) carbon atoms normally requires an antiperiplanar arrangement of the two groups which are to be removed. This has consequences for the rate of reaction (and in extreme cases on whether it proceeds or not) and for the stereochemistry of the product.

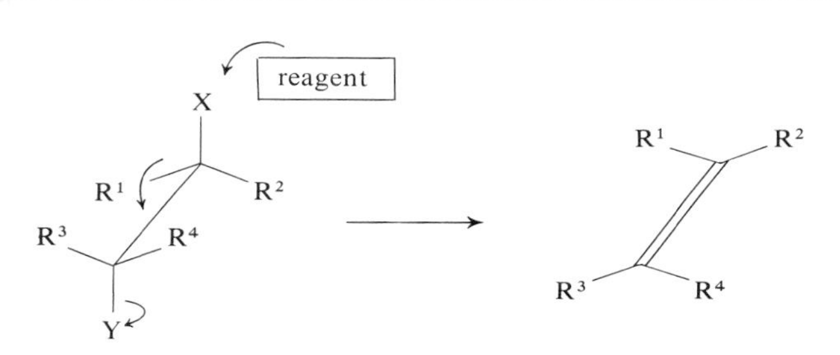

Scheme 2 Elimination reaction showing the antiperiplanar arrangement of the two groups which are to be eliminated.

With acyclic compounds both *threo* and *erythro* isomers can usually assume an *ap* arrangement. These however may vary in stability depending on steric interactions between the remaining groups and this will affect the reaction rates of the two isomers.

Among cyclohexane derivatives *trans*-1,2-disubstituted compounds ($1e2e \gg 1a2a$, see p. 62) can assume an *ap* conformation ($1a2a$) but their *cis*-isomers cannot ($1e2a = 1a2e$, see p. 62). The *trans*-isomers therefore undergo elimination reactions more readily and in many cases the *cis*-isomers do not react at all or undergo an alternative reaction.

In the conversion of alkenes to alkynes only one isomer can assume the necessary *trans* arrangement of groups.

These general principles underly the stereochemistry of the elimination processes which will now be reviewed briefly.

5.5 Dehydrobromination

Dehydrobromination is a *trans*-elimination effected by base and used, for example, in the conversion of dibromides (themselves prepared from alkenes) to alkynes. The bromination–dehydrobromination reaction may be written:

$$RCH{=}CHR' \xrightarrow{Br_2} RCHBrCHBrR' \xrightarrow[-HBr]{base} [RCH{=}CBrR'] \xrightarrow[-HBr]{base} RC{\equiv}CR'$$

There is only one alkyne but two stereoisomeric vinyl bromides and what is known about the addition process (bromination) and the elimination process (dehydrobromination), both of which are *trans*, enables us to decide whether it matters which alkene is used or whether one alkene is superior or perhaps even a requirement for successful synthesis of the alkyne. These possibilities are examined in Schemes **3** and **4**, which should be carefully examined. The conclusions are in line with the observation that *cis*-alkenes readily yield alkynes via *threo* dibromides but that *trans*-alkenes do not. The *erythro* dibromides formed from the latter usually give a mixture of allenes and vinyl halides:

$$\underset{trans}{RCH_2CH{=}CHCH_2R'} \longrightarrow \underset{erythro}{RCH_2CHBrCHBrCH_2R'} \longrightarrow \left\{\begin{array}{l} RCH_2CH{=}CBrCH_2R' \\ RCH_2CBr{=}CHCH_2R' \\ RCH{=}C{=}CHCH_2R' \\ RCH_2CH{=}C{=}CHR' \end{array}\right.$$

5.6 Dehalogenation

Dehalogenation – most often of dibromides – by reaction with zinc or with sodium iodide is a very useful process. The reaction with iodide ion is more highly stereospecific than that with the metal. Among acyclic compounds *threo* dibromides furnish *cis*-alkenes and *erythro* dibromides give *trans*-alkenes with the *erythro* compound slightly more active since it can more readily assume the antiperiplanar conformation. Steric interaction is somewhat greater in the *ap*-configuration of the *threo* isomer. This point is illustrated in Scheme 5 with one *threo* and one *erythro* enantiomer. The reader should satisfy himself that these observations are equally true for the other enantiomers.

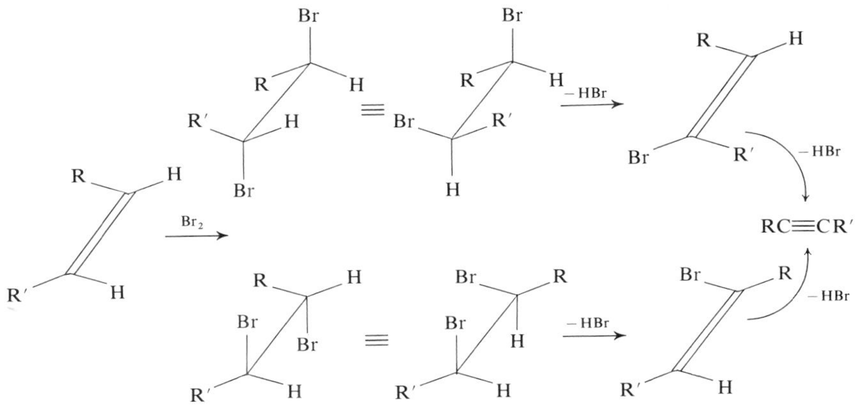

R H
Br₂
R′ H

Scheme 3 Bromination and dehydrobromination of a *cis*-alkene. For the *trans*-elimination of HBr the appropriate groups must assume an antiperiplanar or *trans*-arrangement.

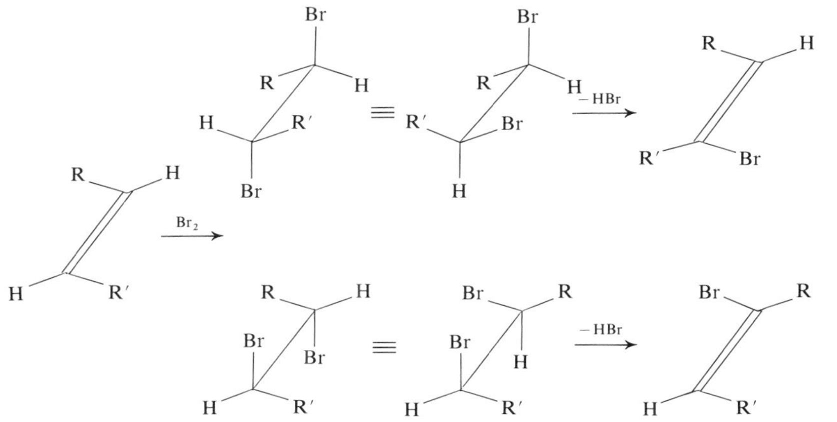

Scheme 4 Bromination and dehydrobromination of a *trans*-alkene. Elimination of one mole of HBr occurs readily from the *erythro* dibromide in an antiperiplanar conformation but the product is a mixture of two vinyl bromides in which bromine and olefinic hydrogen cannot assume a *trans*-arrangement.

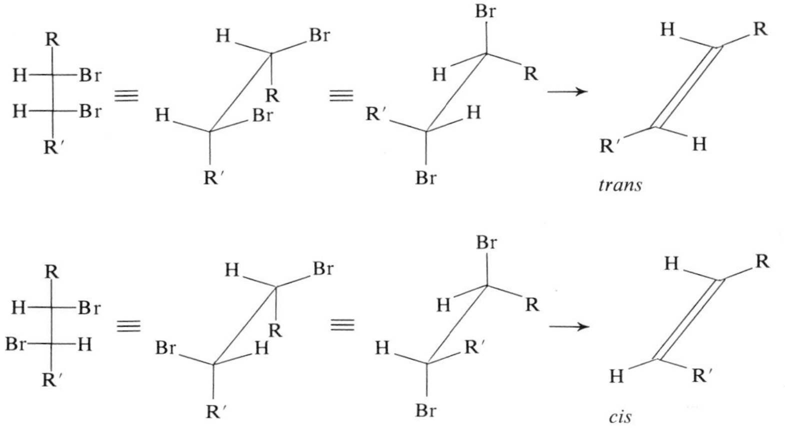

Scheme 5 Debromination of *erythro* and *threo* dibromides.

Q5.4 Stilbene may be brominated and the resulting dibromide reconverted to stilbene thus:

$$PhCH{=}CHPh \longrightarrow PhCHBrCHBrPh \longrightarrow PhCH{=}CHPh$$

Given that both bromination and debromination occur in a *trans*-manner, use sawhorse projections to discover whether the *cis*-alkenes gives the *cis*- or *trans*-isomer as a result of these two reactions.

Q5.5 Bromination of an $\alpha\beta$-unsaturated acid may be accompanied by a second reaction, especially in alkaline solution, involving loss of carbon dioxide and formation of a vinyl bromide.

$$RCH{=}CHCO_2^- \xrightarrow{Br_2} [RCHBrCHBrCO_2^-] \longrightarrow RCH{=}CHBr + CO_2 + Br^-$$

Given that the addition and elimination steps are both *trans* what is the configuration of the vinyl bromide produced from the *trans*-acid?

Compared with acyclic compounds the difference in the reactivity of *cis*- and *trans*-1,2-dibromocyclohexane is more marked since only the *trans*-form can assume the diaxial (antiperiplanar) conformation required to produce cyclohexene. (Cyclohexene exists in only one isomeric form. The smallest ring system in which a *trans* double bond can be accommoda-ted is C_8.) The *cis*-dibromide forms cyclohexene only slowly, probably after substitution (with inversion) of one of the bromine atoms by iodine. These changes are outlined in Scheme 6.

Br Br ≡ Br Br →

1*e*2*e* 1*a*2*a*

Br Br $\xrightarrow{I^-}$ I Br →

1*e*2*a* 1*a*2*a*

Scheme 6 Debromination of *trans*- and *cis*-1,2-dibromocyclohexane (hydrogen atoms omitted for clarity).

The formation of alkenes from quaternary ammonium hydroxides (Hofmann elimination) follows a similar pattern of *trans*-elimination. Reaction occurs smoothly only when the $\overset{+}{N}R_3$ group and a hydrogen atom can assume an *ap*-conformation. This leads to different reaction rates among acyclic compounds and to a restriction of the reaction in some cyclic amines.

$$R'CH_2\underset{\overset{|}{^{+}NR_3\bar{O}H}}{CH}R'' \xrightarrow{\text{heat}} R'CH{=}CHR'' + R_3N + H_2O$$

5.7 *Cis* elimination reactions

Important *cis*-elimination reactions occur through intermediates with some cyclic character. These reactions include the pyrolytic cleavage (300–500°) of esters (acetates and xanthates) and of amine-N-oxides. The reaction occurs through a cyclic intermediate (shown to the right of each equation) in which the groups to be eliminated assume the synperiplanar conformation.

$RCH_2CH(OCOCH_3)R' \longrightarrow RCH{=}CHR'$
acetate

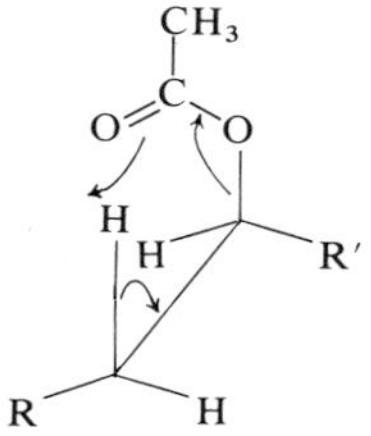

CH_3 C O O H H R' R H

$RCH_2CH(OCSSMe)R' \longrightarrow RCH{=}CHR'$
xanthate

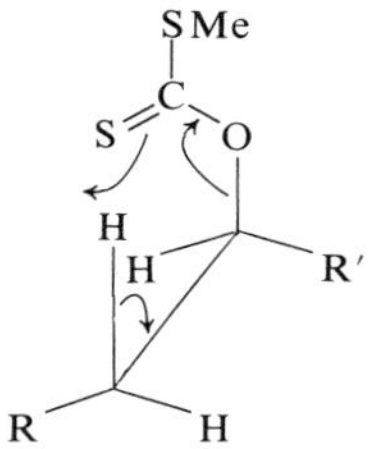

SMe C S O H H R' R H

$R'CH_2CH(\overset{+}{N}\overset{-}{O}R_2)R' \longrightarrow RCH{=}CHR'$
amine oxide

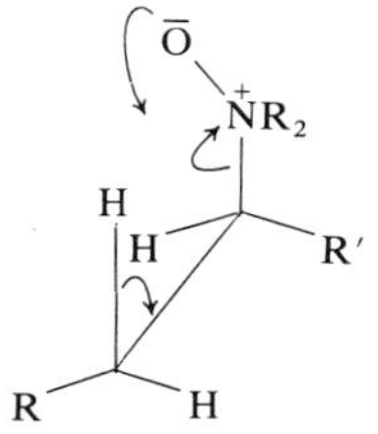

$\bar{O}$ $\overset{+}{N}R_2$ H H R' R H

An interesting example of *cis*-elimination is found in the pyrolysis of the deuterated compound PhCHDCH(OAc)Ph to stilbene. The acetoxy deuterio compound exists as *threo* and *erythro* isomers only one of which gives a product containing deuterium. This observation can be rationalised on the basis that in the preferred conformation for each isomer the bulky phenyl groups will be as far away from one another as possible and the *cis*-elimination then involves AcOH or AcOD as appropriate.

Q5.6 Which of the acetoxy deuterio compounds – the *threo* or the *erythro* isomer – gives the deuterium-containing stilbene?

The Corey–Winter synthesis of alkenes from diols via their thionocarbonates is essentially a *cis*-elimination with the *threo* diol furnishing the *trans*-alkene and the *erythro* diol yielding the *cis*-alkene.

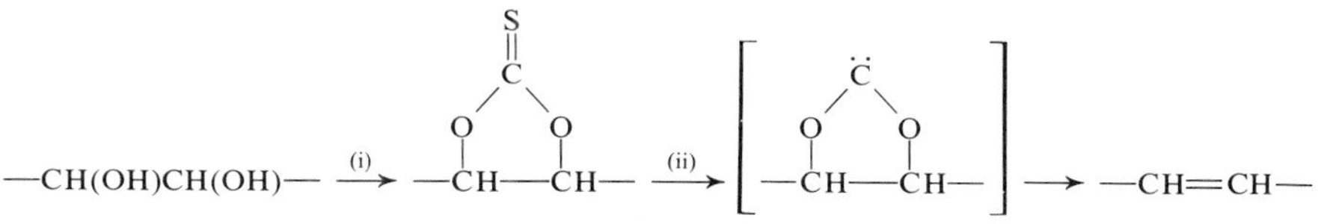

—CH(OH)CH(OH)— $\xrightarrow{(i)}$ —CH—CH— (cyclic thionocarbonate, O–C(=S)–O) $\xrightarrow{(ii)}$ [—CH—CH— (cyclic carbene, O–C̈–O)] ⟶ —CH=CH—

(i) thiocarbonyldi-imidazole, (ii) trimethyl phosphite

5.8 Substitution reactions

It is a useful generalisation, though slightly over-simplified, to say that substitution reactions are accompanied either by inversion or by racemisation. A third possibility – retention of configuration – is relatively uncommon but important when it does occur (Scheme 7). It has been amply

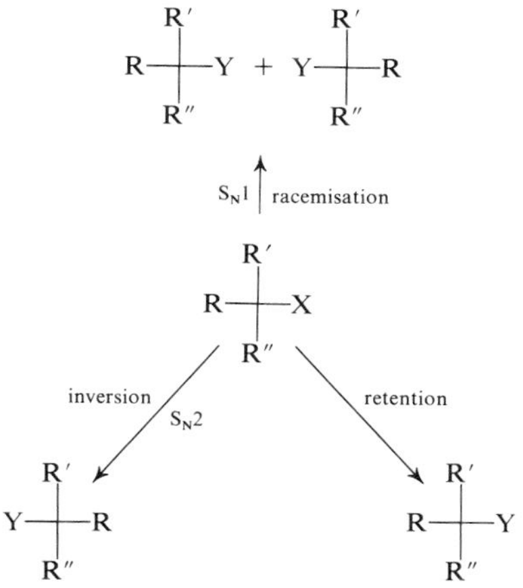

Scheme 7 The possible steric courses of substitution reactions.

demonstrated that bimolecular substitution reactions (S_N2) occur with inversion of configuration and that unimolecular substitution reactions (S_N1) are normally accompanied by racemisation. Retention of configuration is observed in a few S_N1 reactions of compounds containing configuration-holding groups, in S_Ni reactions, and in S_N2 reactions which, on closer examination, are shown to occur through two steps both of which involve inversion.

5.9 Some words of caution

Before elaborating these introductory statements some possible errors must be pointed out. In particular, it must be noted that the change of sign in equation 1 does *not* (in the absence of corroborative evidence) indicate

$$(+)\text{-Cwxyz} \longrightarrow (-)\text{-Cwxya} \qquad \textbf{(Eq. 1)}$$

whether or not there has been a change of configuration. The substrate and product are different compounds and there is no direct relation between sign of rotation and absolute configuration of different compounds.

If the reaction is extended (equation 2) so that the final product is the enantiomer of the substrate then it is clear that the *overall result* is an

$$(+)\text{-Cwxyz} \xrightarrow{\text{(i)}} (-)\text{-Cwxya} \xrightarrow{\text{(ii)}} (-)\text{-Cwxyz} \qquad \textbf{(Eq. 2)}$$

inversion of configuration but there is still insufficient evidence to know whether this occurs at stage (i) or stage (ii). If the reaction proceeds with overall retention of configuration (equation 3) then either there is inversion at both stages or there is no inversion.

$$(+)\text{-Cwxyz} \longrightarrow (-)\text{-Cwxya} \longrightarrow (+)\text{-Cwxyz} \qquad \textbf{(Eq. 3)}$$

The circular arrow in equation 4 is used as a symbol for stereochemical inversion.

$$(R)\text{-A} \xrightarrow{\curvearrowright} \text{B} \qquad \textbf{(Eq. 4)}$$

It should further be realised that the designation of the configuration of B in *R*/*S* terms is not as simple as it may seem. *Two* questions must be asked: does the reaction occur with inversion of configuration and does the change in the molecule lead to a change in the order of precedence of ligands attached to a chiral centre?† Conversely, a change from (*R*) to (*S*) does not immediately indicate a change of configuration – the order of precedence must also be reconsidered.

Q5.7 Assign the appropriate symbol (*R* or *S*) to the products of each of the following reactions:

$$(R)\text{-}CH_3CH(OH)CO_2Me \xrightarrow{\curvearrowright} CH_3CHBrCO_2Me$$

$$(S)\text{-}CH_3CHBrCO_2Me \xrightarrow{\curvearrowright} CH_3CH\begin{matrix} / CH(CO_2Me)_2 \\ \backslash CO_2Me \end{matrix}$$

5.10 Substitution reactions (again)

Current views on the mechanism of substitution reactions have developed from experimental results such as the observation that enantiomeric substrates sometimes furnish optically active products and sometimes racemic products and that changes in experimental conditions may result in a change of reaction order.

Simple substitution (equation 5) involves the breaking of one bond (A—B) and the making of another (B—C). This change is known to happen in two ways depending on the timing of these events. Sometimes the

$$A{-}B + \bar{C} \longrightarrow \bar{A} + B{-}C \qquad \textbf{(Eq. 5)}$$

A—B bond is broken before the B—C bond is made (equation 6). This is a unimolecular reaction with its rate dependent on the concentration of A—B but independent of the concentration of C^- and designated S_N1 (substitution, nucleophilic, unimolecular).

$$(S_N1)\ A{-}B \xrightarrow[\text{step}]{\text{rate-determining}} \bar{A} + \overset{+}{B} \xrightarrow[\text{fast}]{\bar{C}} \bar{A} + B{-}C \qquad \textbf{(Eq. 6)}$$

† A change in one of the ligands at a point some distance from the chiral centre, such as hydrogenation of a double bond, may be sufficient to change the *R*/*S* symbols and this feature is considered by some to be a disadvantage of this system.

Other substitution reactions are *bimolecular* and proceed at a rate depending on the concentration of A—B *and* of C. These are designated S_N2 (substitution, nucleophilic, bimolecular) and are believed to proceed

$$(S_N2)\ \text{A-B} + \text{C} \xrightarrow[\text{step}]{\text{rate-determining}} \text{A---B---C} \longrightarrow \text{A} + \text{B-C} \qquad \textbf{(Eq. 7)}$$

through an intermediate of the type shown in equation 7. Bond-making and bond-breaking proceed simultaneously.

S_N2 reactions always occur with inversion of configuration: S_N1 reactions usually occur with racemisation though this is not always complete. In reality the position is not always as clear cut as this because the S_N1 and S_N2 reactions are now recognised as limiting situations which reactions approach but may not fully attain.

The inversion which always accompanies S_N2 reactions can be understood in terms of the nature of the intermediate in which the incoming group (reagent) approaches the reaction centre from the opposite side to the leaving group. (This is sometimes described as a Walden inversion after the chemist who first demonstrated, in 1896, that inversion must occur in some reactions, p. 93). It is often compared with an umbrella being blown inside out by the wind.

$\bar{\text{Y}}$ + (A)(B)(C)C—X ⟶ Y---C(A)(B)(C)---X ⟶ Y—C(A)(B)(C) + $\bar{\text{X}}$

In an S_N1 reaction the intermediate is a carbenium (carbonium) ion which, given a sufficiently long life, will assume a planar configuration and be subject to subsequent attack by the reagent equally from either side. This simple consideration is appropriate with stable carbenium ions. With less stable cations the leaving group $\bar{\text{X}}$ may remain close enough to cause

(A)(B)(C)C—X ⟶ (A)(B)(C)C$^+$ ⟶ (A)(B)(C)C—Y + Y—C(A)(B)(C)

enantiomeric substrate — planar carbenium ion — racemic product

the incoming group to attack preferentially – but not exclusively – from the opposite side. The result is partial inversion and partial racemisation as in the following examples of S_N1 reactions:

$$PhCHClCH_3 \xrightarrow[\text{aq. acetone}]{\text{hydrolysis}} PhCH(OH)CH_3 \qquad \text{90 per cent racemisation}$$

$$C_6H_{13}CHBrCH_3 \xrightarrow{\text{hydrolysis}} C_6H_{13}CH(OH)CH_3 \qquad \text{30 per cent racemisation}$$

Retention of configuration is fairly uncommon. It is observed, however, in S_N1 reactions where there is a neighbouring group able to interact with the developing carbenium ion centre, and to hold this in its normal tetrahedral form. This is true, for example, of the CO_2H group and the following case is typical.

Hydrolysis of 2-bromopropanoic acid with concentrated alkali occurs by the S_N2 process and with the expected stereochemical result – inversion. With dilute alkali, however, the reaction becomes unimolecular and because of the influence of the adjacent carboxyl group reaction occurs with retention of configuration. This example illustrates the practical

$$CH_3CHBrCO_2H \xrightarrow[S_N2]{\text{conc KOH}} CH_3CH(OH)CO_2H \qquad \text{inversion}$$

$$CH_3CHBrCO_2H \xrightarrow[S_N1]{\text{dil KOH}} CH_3CH(OH)CO_2H \qquad \text{retention}$$

difficulty of maintaining optical purity during chemical reactions since a change of reaction conditions – in this case of concentration of reagent – leads to a different mechanism and to a change in the stereochemical result.

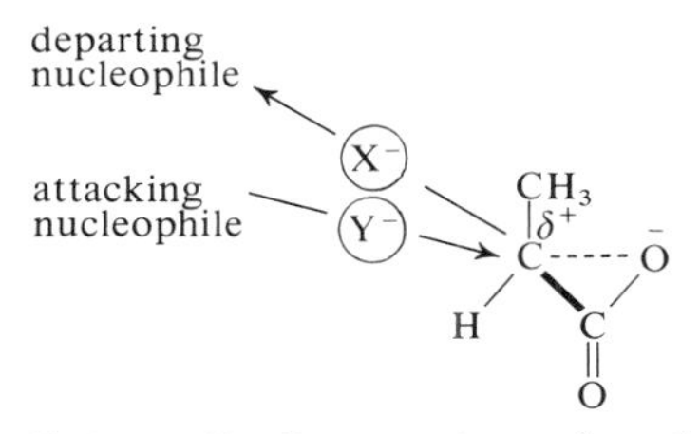

Scheme 8 Interaction of carboxylate anion with the carbenium centre holds the reactive intermediate in its tetrahedral form and requires the attacking nucleophile to approach from the same side as the departing group.

A typical S_Ni (substitution, nucleophilic, internal) reaction is the chlorination of enantiomeric alcohols with thionyl chloride which occurs with retention of configuration and has been formulated as a two-stage process in the following terms:

$$R^1R^2R^3C{-}OH \xrightarrow{SOCl_2} R^1R^2R^3C{-}O{-}S(=O){-}Cl \longrightarrow R^1R^2R^3C{-}Cl + SO_2$$

More recent studies suggest that the second stage is not concerted but proceeds through an ion pair.

5.11 Two further points

The stereochemical fate of a molecule containing two chiral centres depends on whether reaction occurs at one or both centres. If reaction occurs at only one centre and is accompanied by inversion then a *threo* enantiomer will become an *erythro* enantiomer and a *threo* racemate will furnish the *erythro* racemate. The change of configuration at one chiral centre in a molecule containing more than one such centre is called **epimerisation**. Such a change leads to an interconversion of diastereo-isomers. The situation occurring when reaction occurs at both centres with inversion is developed in Q5.8.

Q5.8 For the reaction summarised in the following equation

$$\text{RCHXCHXR}' \xrightarrow{S_N2} \text{RCHYCHYR}'$$

what product(s) result(s) from:

(i)

```
      R
  H---+---X
  H---+---X
      R'
```

(ii)

```
      R
  X---+---H
  X---+---H
      R'
```

(iii) the *erythro* racemate?

Reactions which do not involve the bond by which a ligand is attached to a chiral centre would not be expected to affect the configuration of the molecule. For example, the acetylation of the alcohol in equation 8 does

$$\text{R}-\underset{\text{R}''}{\overset{\text{R}'}{\text{C}}}-\text{OH} \underset{\text{hydrolysis}}{\overset{\text{acetylation}}{\rightleftharpoons}} \text{R}-\underset{\text{R}''}{\overset{\text{R}'}{\text{C}}}-\text{OAc} \qquad \textbf{(Eq. 8)}$$

not involve the C—O bond and reaction should occur without any change of configuration. Similarly, under normal conditions ester hydrolysis does not affect the (alkyl) C—O bond and this reaction should also occur without change of configuration. This principle can be used as part of a procedure for interconverting enantiomeric alcohols (see p. 94).

5.12 Some interesting examples

The general principles which have been outlined can now be applied to a number of interesting examples.

(i) When Walden (1896) reported the reactions shown in Scheme 9 he clearly demonstrated, for the first time, the possibility of inversion, for up to that time it was accepted that all substitution reactions occurred without inversion of configuration. For example, (+)-chlorosuccinic acid is converted to malic acid in two ways: with aqueous silver oxide the (+)-enantiomer is formed, and with potassium hydroxide the (−)-enantiomer.

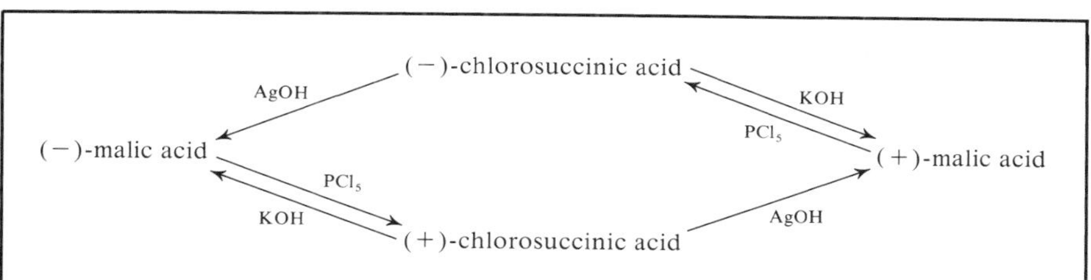

Scheme 9 Interconversions of malic acid [$HO_2CCH(OH)CH_2CO_2H$] and chlorosuccinic acid [$HO_2CCHClCH_2CO_2H$].

One of these reactions must occur with inversion and the other with retention of configuration. It is now known that inversion is the rule rather than the exception and that all the reactions shown in Scheme 9 occur with inversion except those with aqueous silver oxide. It follows that enantiomers of malic acid and chlorosuccinic acid with the same sign of rotation also have the same configuration.

(ii) Ethyl (+)-lactate is converted to the acetoxy ester by two routes which furnish the (+) and (−) enantiomers respectively and this sequence is easily interpreted in terms of the earlier discussion. The conversion of

$$\begin{array}{ccc} (+)\text{-}CH_3CH(OH)CO_2Et & \xrightarrow{TsCl^*} & (+)\text{-}CH_3CH(OTs)CO_2Et \\ \downarrow{\scriptstyle AcCl} & & \downarrow{\scriptstyle KOAc} \\ (+)\text{-}CH_3CH(OAc)CO_2Et & & (-)\text{-}CH_3CH(OAc)CO_2Et \end{array}$$

* TsCl = *p*-toluenesulphonyl chloride

the hydroxyl group to its acetate by reaction with acetyl chloride or to its toluenesulphonate by reaction with toluenesulphonyl chloride do not involve the chiral C—O bond and should therefore occur without change of configuration. The conversion of tosylate to acetate, however, is an S_N2 process in which $\overline{O}$Ac displaces $\overline{O}$Ts and would be expected to occur with inversion. Hence the lactate and acetate with the same sign of rotation also have the same configuration. By a similar argument the deacetylation of acetoxypropionate should occur without change of configuration.

$$(-)\text{-}CH_3CH(OAc)CO_2Et \xrightarrow[\text{NaOEt, EtOH}]{\text{EtOH, H}^+\text{ or}} (-)\text{-}CH_3CH(OH)CO_2Et$$

(iii) Hydrogenation of a natural unsaturated acid gives 9-hydroxystearic acid which has been shown to have the *R* configuration. This compound has been used as a source of an enantiomeric 9-tritiostearic acid (equation 9) required for the study of the enzymic desaturation process by

$$>CH(OH) \xrightarrow{\text{TsCl}} >CH(OTs) \xrightarrow{\text{LiAlT}_4} >CHT \qquad \textbf{(Eq. 9)}$$

which stearic acid is converted to oleic acid. The enantiomeric tritiostearic acid is similarly prepared from the enantiomeric hydroxy acid (see Q9 and 10).

(iv) *Cis* epoxides can be used as a source of both *cis*- and *trans*-alkenes by application of appropriate sequences of stereospecific reactions detailed in Q11.

Q5.9 From the information given along with equation 9 deduce the configuration of the 9-tritiostearic acid produced from (*R*)-9-hydroxystearic acid.

Q5.10 The hydroxystearic acid required for the (*R*)-9-tritiostearic acid is not readily available. Suggest how (*S*)-9-hydroxystearic acid may be made from its more readily available enantiomer.

Q5.11 Deduce the configuration of the alkene (*cis* or *trans*) resulting from each of the following sequences set out below:

(i)

$$\underset{cis}{\overset{O}{RCHCHR'}} \xrightarrow{H_2O,\ H^+} RCH(OH)CH(OH)R' \xrightarrow{\text{MsCl}\dagger}$$

$$RCH(OMs)CH(OMs)R' \xrightarrow[\textit{trans}\text{ elimination}]{I^-} RCH{=}CHR'$$

† MsCl = methanesulphonyl chloride (CH_3SO_2Cl).

(ii) RCHCHR′ (epoxide, O bridging the two C) *cis* $\xrightarrow{\text{HBr}}$ RCH(OH)CHBrR′ / RCHBrCH(OH)R′ $\xrightarrow{\text{Ac}_2\text{O}}$

RCH(OAc)CHBrR′ / RCHBrCH(OAc)R′ $\xrightarrow{\text{AgOAc, AcOH aq}}$ RCH(OAc)CH(OAc)R′

$\xrightarrow{\text{KOH}}$ RCH(OH)CH(OH)R′ $\xrightarrow{\text{MsCl†}}$ RCH(OMs)CH(OMs)R′

$\xrightarrow[\textit{trans}\text{ elimination}]{\text{I}^-}$ RCH═CHR′

5.13 Neighbouring group participation

A reaction such as that shown in equation 10 in which group B is being replaced by group C may be influenced by group A in a number of ways.

$$\text{A—X—Y—B} \xrightarrow{\text{C}} \text{A—X—Y—C} \quad \textbf{(Eq. 10)}$$

This occurs most frequently when A is attached to X but sometimes also when A is even further removed from the reacting centre Y—B. The general phenomenon is known as neighbouring group participation and sometimes has stereochemical consequences. Participation is possible when there is an interaction between the atom or group A and the atom Y which may become electron-rich or electron-deficient during the breaking of the Y—B bond. Interaction may be apparent in several ways: an enhanced rate of reaction, an unexpected product which does not include the reagent C (**7**), a rearrangement of groups (**8**), or a product of unexpected

(**7**) A bridging X—Y (three-membered ring)

(**8**) C—X—Y—A

stereochemistry. Many groups can be involved in neighbouring group participation with an adjacent electron-deficient reaction centre including the halogens; sulphur, oxygen, and nitrogen-containing groups; carboxylic acids, esters, and amides; and phenyl groups and double bonds. A few examples are discussed.

(i) When an α-hydroxybromide is treated with base the major product is usually the epoxide resulting from intramolecular reaction rather than

† MsCl = methanesulphonyl chloride (CH_3SO_2Cl).

the diol. The epoxide formation (like the bromine hydrolysis had it occurred) is likely to be an S_N2 process and to occur with inversion. Under appropriate reaction conditions the epoxide might be subject to subsequent attack by $\bar{O}H$ but this would probably occur at either of the epoxide carbon atoms – again by an S_N2 reaction and involve a second inversion.

$$RCH(OH)CHBrR' \xrightarrow{\bar{O}H} RCH(\bar{O})CHBrR' \longrightarrow R\overset{O}{CH{-}CH}R' \text{ (epoxide)}$$

$$RCH(OH)CHBrR' \xrightarrow{\bar{O}H} RCH(OH)CH(OH)R'$$

(ii) The sulphur compound in equation 11 is hydrolysed 10,000 times quicker than its O-analogue under the same conditions. This difference probably results from the fact that the cyclic reaction intermediate is formed more readily with sulphur than with the less electronegative oxygen atom.

$$EtS{-}CH_2{-}CH_2{-}Cl \xrightarrow[\text{step}]{\text{rate-determining}} Et\overset{+}{S}{<}(CH_2{-}CH_2) \xrightarrow{\text{fast}} EtS{-}CH_2{-}CH_2{-}OH$$

(Eq. (11)

(iii) Basic hydrolysis of the amino chloride (equation 12) occurs with rearrangement of the main carbon skeleton which is best understood in terms of neighbouring group participation.

$$Et_2N{-}CH_2{-}CHEt{-}Cl \longrightarrow Et_2N{<}(CH(Et){-}CH_2) \xrightarrow{H\bar{O}} HOCH_2CH(Et)NEt_2$$

(Eq. 12)

(iv) In the conversion of α-acetoxy halides to acetylated diols by reaction with silver acetate and acetic acid reaction occurs with retention of configuration under anhydrous conditions but with complete inversion when the solvent contains at least one mole of water. Under anhydrous conditions the acetate ion enters from the same side as the departing halide ion because the acetoxy group interacts with the developing carbenium ion and blocks entry from the opposite side. In the presence of water this reagent interacts with the intermediate in such a manner that OH is introduced from the same side as the acetate group, i.e. the opposite side from the departing halide ion.

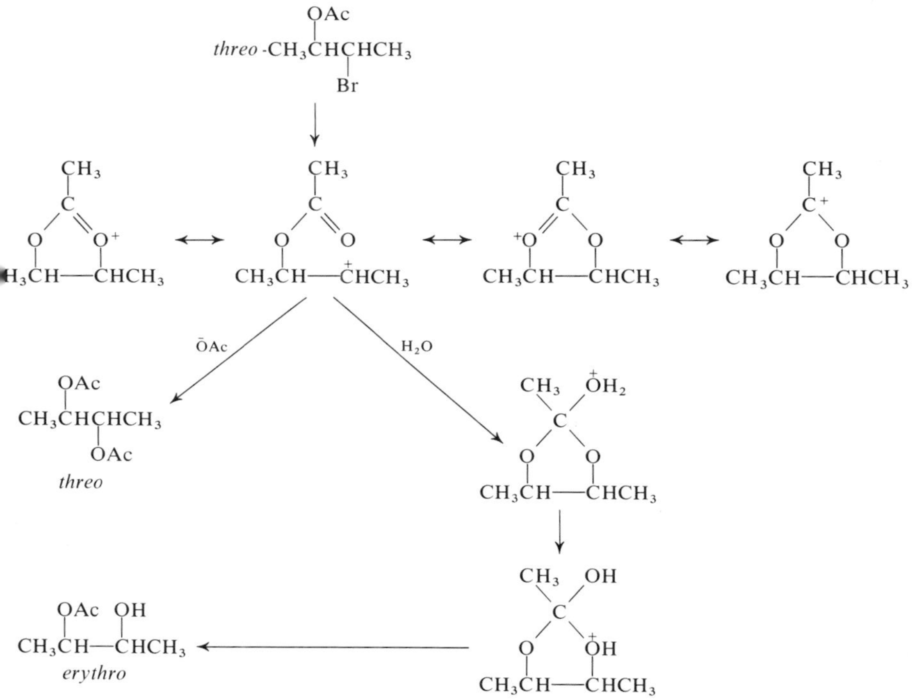

threo-$CH_3CH(OAc)CH(Br)CH_3$

CH_3–C=O$^+$ / O (ring) CH_3CH—$CHCH_3$ ⟷ CH_3–C(=O)O CH_3CH—$\overset{+}{C}HCH_3$ ⟷ CH_3–C(=$^+$O)O CH_3CH—$CHCH_3$ ⟷ CH_3–C^+(O)(O) CH_3CH—$CHCH_3$

$\bar{O}Ac$ → threo $CH_3CH(OAc)CH(OAc)CH_3$

H_2O → CH_3, $\overset{+}{O}H_2$–C(O)(O) CH_3CH—$CHCH_3$ → CH_3, OH–C(O)($\overset{+}{O}H$) CH_3CH—$CHCH_3$ → erythro $CH_3CH(OAc)$—$CH(OH)CH_3$

Scheme 10 Reaction of α-acetoxyhalides with silver acetate and acetic acid in the absence and presence of water.

5.14 Molecular rearrangements

In most reactions of organic compounds one functional group is converted to another without any change in the carbon skeleton of the substrate. Those reactions in which this is not true are described as rearrangements. Two familiar examples are formulated in Schemes 11 and 12. There are several interesting stereochemical problems which may arise in rearrangement reactions and three of them are discussed here.

(i) In rearrangements of the type shown in equation 13 in which the

$$\overset{R}{X}\text{—}Y \longrightarrow X\text{—}\overset{R}{Y} \qquad \textbf{(Eq. 13)}$$

$$Me_3CCH_2OH \xrightarrow{HBr} Me_3C\overset{+}{C}H_2 \longrightarrow Me_2\overset{+}{C}CH_2Me \longrightarrow Me_2CBrCH_2Me$$

primary carbenium ion — tertiary carbenium ion

Scheme 11 The reaction of neopentyl alcohol with HBr giving tert amyl bromide involves the rearrangement of the carbon atoms in the substrate. This is believed to occur because the reaction proceeds through the primary carbenium ion which rearranges to the more stable tertiary carbenium ion.

$$Ph_2C(OH)-C(OH)Ph_2 \overset{H^+}{\rightleftharpoons} Ph_2C(OH)-C(\overset{+}{O}H_2)Ph_2 \rightleftharpoons Ph_2C(OH)-\overset{+}{C}Ph_2 \rightleftharpoons Ph-\overset{+}{C}(OH)-CPh_3 \longrightarrow Ph-C(=O)-CPh_3$$

Scheme 12 The pinacol rearrangement: 1,2-diols (especially ditertiary diols) furnish rearranged ketones when treated with acid by the mechanism shown above. This is represented in discrete steps for the sake of clarity but may follow a concerted mechanism.

migrating group R is attached to X through a chiral centre migration may occur with retention of configuration or with racemisation. When reaction occurs with retention of configuration this is taken as evidence of an

$$\overset{R}{\underset{|}{X}}-Y \rightleftharpoons X \overset{R}{\cdots} Y \rightleftharpoons X-\overset{R}{\underset{|}{Y}}$$

intramolecular reaction in which the R group is never separated from the molecule; bond-making and bond-breaking occur simultaneously. Racemisation, on the other hand, is associated with intermolecular reactions in which the group R separates from its molecule – most often as a carbenium ion which assumes a planar configuration – and loses its stereochemical identity.

Migration with retention of configuration has been observed in many rearrangement reactions including those associated with the names of Hofmann, Curtius, Lossen, Wolff, Beckmann, and Baeyer and Villiger. The conversion of (+)-3-phenylbutan-2-one to (−)-1-phenylethyl acetate by the last-named oxidation is typical (Scheme 13).

$$RR'C{=}O \xrightarrow{PhCO_3H} RR'C(OH)(O{-}OCOPh) \longrightarrow RR'C(O{-}H)(\overset{+}{O}) \longrightarrow R\overset{+}{C}(O{-}H)(OR') \longrightarrow R{-}C({=}O)OR'$$

$R = CH_3$ $\quad R' = $ Ph(CH$_3$)CH—

Scheme 13 Baeyer–Villiger oxidation.

Migration of R′ from carbon to electron-deficient oxygen occurs without any change of configuration of its chiral centre.

(ii) The migration of R from X to Y (equation 13) is usually preceded or followed by the elimination or displacement of some group attached to Y. If Y has four different ligands attached to it then migration will be accompanied by inversion at this centre if this step is bimolecular but by racemisation if the group attached to Y separates first and leaves a carbenium ion which is sufficiently stable to assume a planar configuration.

(iii) In rearrangement reactions where X and Y are multiply bonded the migrating group usually comes from the side opposite to the leaving group. This has been clearly demonstrated in the Beckmann rearrangement of oximes as shown in Scheme 14. It is the group R (on the opposite side of the C=N to the leaving group) which migrates and not R′ as was once thought.

$$\mathrm{R(R')C{=}N{-}OH} \xrightarrow{H^+} \mathrm{R(R')C{=}N{-}\overset{+}{O}H_2} \longrightarrow \mathrm{R'\overset{+}{C}{\equiv}N{-}R} \xrightarrow{H_2O} \mathrm{H_2\overset{+}{O}(R')C{=}N{-}R} \longrightarrow \mathrm{HO(R')C{=}N{-}R} \longrightarrow \mathrm{R'CONHR}$$

Scheme 14 Beckmann rearrangement of a ketoxime under the influence of acid. The migrating group (R) is on the opposite side of the double bond from the leaving group (OH).

5.15 Asymmetric synthesis

The term asymmetric synthesis is applied to reactions which produce two enantiomers or diastereoisomers *in unequal amounts*. This does not happen in the absence of a chiral influence so that, for example, the reduction of 2-oxopropanoic acid (pyruvic) normally gives the racemic form of 2-hydroxypropanoic acid (lactic), the two enantiomeric forms of the hydroxy acid being formed in equal amounts. In an asymmetric synthesis the two enantiomers are formed unequally and this situation may arise with an

$$\underset{\text{pyruvic acid}}{CH_3COCO_2H} \xrightarrow{NaBH_4} \underset{\text{lactic acid}}{(\pm)\text{-}CH_3CH(OH)CO_2H}$$

enantiomeric substrate (for example, the pyruvic acid can be esterified with an enantiomeric alcohol), with an enantiomeric reagent, or even in a reaction conducted in an enantiomeric solvent or under the agency of an enantiomeric catalyst. Asymmetric syntheses have also been observed in photo reactions occurring under the influence of circularly polarised light. This may be one way (of several) by which enantiomeric molecules were first produced. In the extreme case only one enantiomer will be formed. This is hardly ever achieved in ordinary chemical reactions but is normal for most enzymic reactions (enzymes are enantiomeric compounds acting either as substrates or as catalysts). Some examples of asymmetric synthesis will be discussed.

(i) A simple example involves the conversion of aldehyde to hydroxy acid via the cyanhydrin, a reaction in which a new chiral centre is formed. Starting with an achiral aldehyde the hydroxy acid will be racemic but

$$\mathrm{RCHO} \xrightarrow{\mathrm{HCN}} \mathrm{RCH(OH)CN} \xrightarrow{\text{hydrolysis}} \mathrm{RCH(OH)CO_2H}$$

with an enantiomeric aldehyde the product will probably be a mixture of two diastereoisomers. The intermediate or transition state through which these are formed and the diastereoisomeric products themselves need not be equally stable and therefore need not be formed in equal amounts. The conversion of (+)-arabinose to gluconic and mannonic acids is known to give these two acids in a 1:3 ratio.

arabinose		gluconic acid		mannonic acid
CHO		CO_2H		CO_2H
H—OH		HO—H		H—OH
HO—H	→	H—OH	+	H—OH
HO—H		HO—H		HO—H
CH_2OH		HO—H		HO—H
		CH_2OH		CH_2OH

(ii) Reduction of the α-oxo acid 2-phenyl-2-oxo-ethanoic acid gives the racemic hydroxy acid but reduction of its ester with (−)-menthol gives predominantly the ester of (−)-mandelic acid (Scheme 15).

$$\mathrm{PhCOCO_2H} \xrightarrow{\text{esterification}} \underset{(-)\text{-menthyl ester}}{\mathrm{PhCOCO_2R}} \xrightarrow{\text{reduction}}$$

$$\mathrm{PhCH(OH)CO_2R} \xrightarrow{\text{hydrolysis}} \underset{\text{mandelic acid}}{\mathrm{PhCH(OH)CO_2H}}$$

Scheme 15 Asymmetric synthesis of mandelic acid by reduction of the (−)-menthyl ester of the oxo acid.

(iii) Asymmetric analysis is also observed in the reactions of Grignard reagents with enantiomeric ketones or with esters of α-oxo acids with enantiomeric alcohols. These results have been rationalised in terms of the

$$\mathrm{RCOCO_2H} \longrightarrow \mathrm{RCOCO_2R'} \xrightarrow{\mathrm{MeMgI}} \mathrm{C(R)(Me)(HO)(CO_2R')} \longrightarrow \mathrm{C(R)(Me)(HO)(CO_2H)}$$

preferred conformation of the substrate (Cram's rule) and if the absolute configuration of the esterifying alcohol is known then the absolute configuration of the α-hydroxy acid formed in excess can be deduced. This provides a useful method of determining absolute configuration and is detailed in Scheme 16.

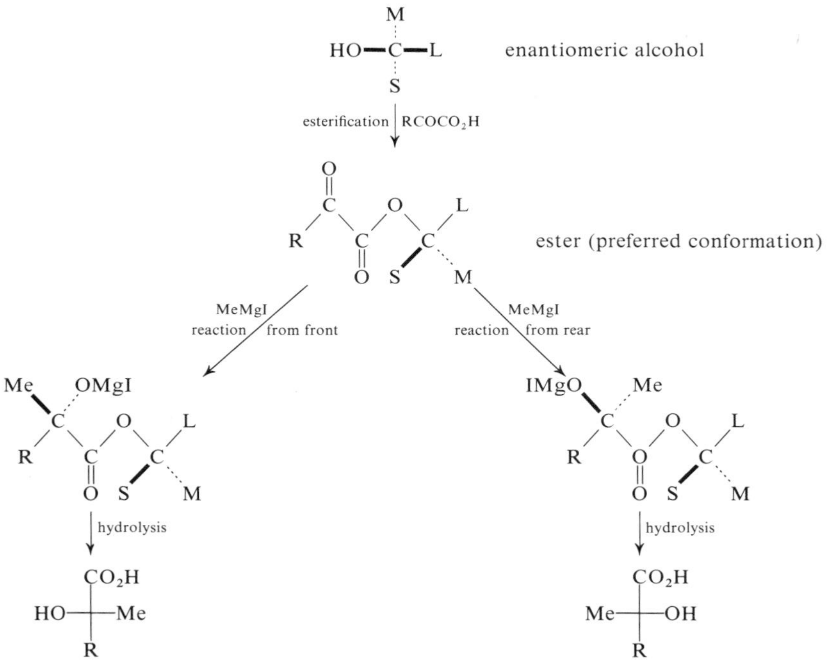

M
HO—C—L enantiomeric alcohol
S

esterification | $RCOCO_2H$

ester (preferred conformation)

MeMgI reaction from front

MeMgI reaction from rear

hydrolysis

CO_2H
HO—|—Me
R

CO_2H
Me—|—OH
R

Scheme 16 The enantiomeric alcohol (with large, medium, and small ligands) forms an ester with the preferred conformation shown. The result of reaction from the front and from the back of the carbonyl group is shown on the left and the right respectively. Attack is more likely to occur from the less hindered side (i.e. the front where the ligand is S rather than the back where the ligand is M) so that the hydrolysed acid will contain an excess of that shown on the left. A similar sequence can be used to show that the enantiomeric alcohol would furnish an excess of the enantiomeric hydroxy acid.

(iv) Hydration of an alkene via hydroboration gives a product with a new chiral centre which will normally be formed as a racemic mixture.

$$RCH{=}CHR' \xrightarrow{(BH_3)_2} \left[RCH_2{-}CH\begin{matrix} BH_2 \\ R' \end{matrix} \right] \xrightarrow{H_2O_2} RCH_2CH(OH)R'$$

Asymmetric synthesis has been observed when the diborane is replaced by an enantiomeric dialkylborane (R''_2BH) and it is claimed that *cis*-but-2-ene gives almost pure (−)-butan-2-ol by reaction with the dialkylated borane from an enantiomer of α-pinene.

(v) Reactions are likely to be influenced by enantiomeric solvents only when solvation of reactants or intermediates is an important feature of the process. The reaction between benzaldehyde and ethylmagnesium bromide gives 1-phenylpropan-1-ol which is optically active when conducted in the presence of (+)-2,3-dimethoxybutane. The enantiomeric ether is part of the Grignard complex which reacts with the aldehyde.

(vi) The influence of enantiomeric catalysts has been demonstrated in the preferential formation of an enantiomeric cyanohydrin in the presence of an enantiomeric base such as quinine.

$$PhCH{=}CHCHO + HCN \xrightarrow[\text{quinine}]{} PhCH{=}CHCH(OH)CN$$

Answers

A5.1

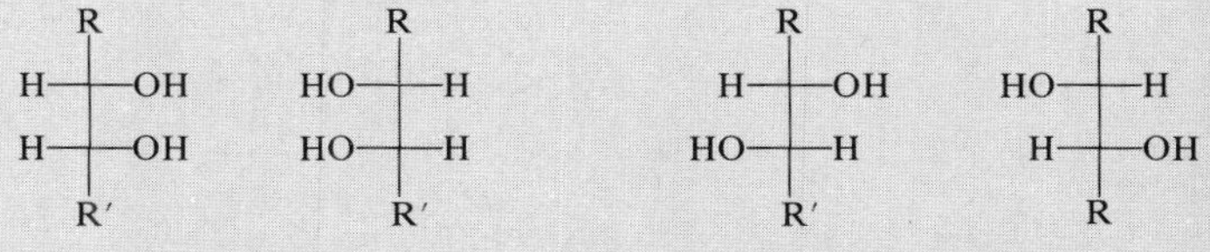

$$\begin{matrix} R & & R' \\ & C{=}C & \\ H & & H \end{matrix} \quad (Z) \qquad \begin{matrix} R & & H \\ & C{=}C & \\ H & & R \end{matrix} \quad (E)$$

$$\begin{matrix} & R & \\ H- & + & -OH \\ H- & + & -OH \\ & R' & \end{matrix} \qquad \begin{matrix} & R & \\ HO- & + & -H \\ HO- & + & -H \\ & R' & \end{matrix} \qquad \begin{matrix} & R & \\ H- & + & -OH \\ HO- & + & -H \\ & R' & \end{matrix} \qquad \begin{matrix} & R & \\ HO- & + & -H \\ H- & + & -OH \\ & R & \end{matrix}$$

erythro *threo*

A5.4

Ph H
Ph H
cis

Br_2 →

Br
Ph H
Ph H
Br

Ph H
Br
Br
Ph H
threo

$-Br_2$ →

$-Br_2$ →

Ph H
Ph H
cis

A5.5

H CO_2^-
R H
trans

Br_2 →

Br
H CO_2^-
R H
Br

≡

O O
C
Br H
R H
Br

H CO_2^-
Br
Br
R H

≡

Br H
Br
C
O O^-
R H

→

Br H
R H
cis

The elimination step involves loss of CO_2 and Br^- and requires a *trans*-arrangement of these two groups.

A5.6

Ph
H—D
H—OAc
Ph
erythro

≡

D Ph
H
Ph H
OAc

→

D Ph
Ph H
trans-stilbene

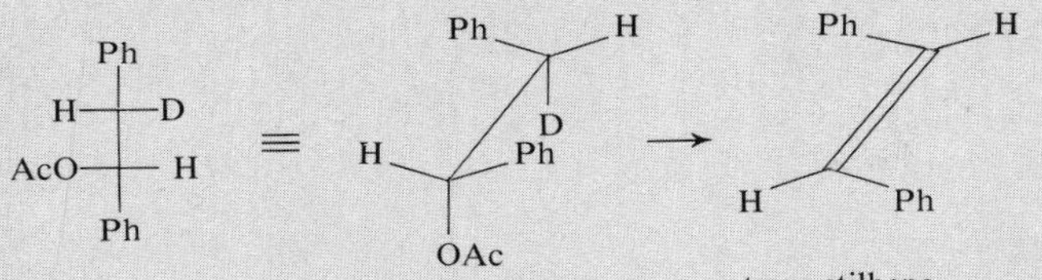

threo   *trans*-stilbene

Both isomers give *trans*-stilbene: the *threo* isomer furnishes a D-free compound and the *erythro* isomer gives a D-containing product.

A5.7 (i) *S* inversion and *no* change in the order of precedence.
(ii) *S* inversion and a change in the order of precedence.

A5.8 (i) (ii) 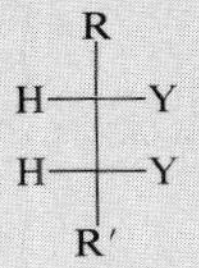 (iii) the *erythro* racemate [i.e. (i) and (ii)]

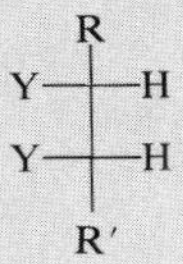

since inversion occurs at both centres the enantiomeric substrates (i) and (ii) furnish enantiomeric products.

A5.9 *S*

A5.10 By the following sequence of reactions which involves one inversion:

>CH(OH) ⟶ >CH(OTs) ⟶ >CH(OAc) ⟶ >CH(OH)

A5.11 *cis* ⟶ *threo* ⟶ *threo* ⟶ *cis*

cis ⟶ *threo* ⟶ *threo* ⟶ *erythro* ⟶ *erythro*

⟶ *erythro* ⟶ *trans*

References

Anon., *J. Org. Chem.*, 1970, **35**, 2849.

J. M. Bijvoet, A. F. Peerdeman, and A. J. van Bommell, *Nature*, 1951, **168**, 271.

J. H. Brewster in *Techniques of Chemistry, Vol. IV, Elucidation of Organic Structures by Physical and Chemical Methods (2nd edn)*, pp. 6–38, ed. K. W. Bentley and G. W. Kirby, Wiley Interscience, New York, 1972.

R. S. Cahn and C. K. Ingold, *J. Chem. Soc.*, 1951, 612.

R. S. Cahn, C. K. Ingold, and V. Prelog, *Experientia*, 1956, **12**, 81.

R. S. Cahn, C. K. Ingold, and V. Prelog, *Angew. Chem. internat. edit.*, 1966, **5**, 385.

E. L. Eliel, *J. Chem. Educ.*, 1971, **48**, 163.

Von Meyer and Jacobsen, *Lehrbuch der Organische Chemie*, 1893 p. 513, Beit & Co., Leipzig.

J. Read and F. D. Gunstone, *Text-book of Organic Chemistry*, 4th edn, 1958, p. 271, Bell, London.

Additional References

Further information on the topics covered in this book is available in the following references:

Progress in Stereochemistry, Butterworths, London
Vol. **1**. Ed. W. Klyne, 1954.
Vol. **2**. Ed. W. Klyne and P. B. D. de la Mare, 1958.
Vol. **3**. Ed. P. B. D. de la Mare and W. Klyne, 1962.
Vol. **4**. Ed. B. J. Aylett and M. M. Harris, 1969.

Topics in Stereochemistry, Wiley, New York
Ed. N. L. Allinger and E. L. Eliel: Vol. **1**, 1967; **2**, 1968; **3**, 1969; **4**, 1970; **5**, 1971; **6**, 1972; **7**, 1973; **8**, 1974.

E. L. Eliel, *Stereochemistry of Carbon Compounds*, McGraw-Hill, London, 1962.

J. Grundy, *Stereochemistry, The Static Principles*, Butterworths, London, 1964.

K. Mislow, *Introduction to Stereochemistry*, Benjamin, New York, 1965.

E. L. Eliel, N. L. Allinger, S. J. Angyal, and G. A. Morrison, *Conformational Analysis*, Interscience, New York, 1965.

G. Hallas, *Organic Stereochemistry*, McGraw-Hill, London, 1965.

E. L. Eliel, *Elements of Stereochemistry*, Wiley, New York, 1969.

K. Bláha, O. Červinka, and J. Kovár, *Fundamentals of Stereochemistry and Conformational Analysis*, Iliffe Books, London, 1971.

G. Natta and M. Farina, *Stereochemistry*, Longman, London, 1972.

D. Whittaker, *Stereochemistry and Mechanism*, Oxford, 1973.

F. D. Gunstone, *Basic Stereochemistry*, Programmes in Organic Chemistry, Volume 8, English Universities Press, 1974.

Index